AF760339

RILEM State-of-the-Art Reports

RILEM STATE-OF-THE-ART REPORTS
Volume 45

RILEM, The International Union of Laboratories and Experts in Construction Materials, Systems and Structures, founded in 1947, is a non-governmental scientific association whose goal is to contribute to progress in the construction sciences, techniques and industries, essentially by means of the communication it fosters between research and practice. RILEM's focus is on construction materials and their use in building and civil engineering structures, covering all phases of the building process from manufacture to use and recycling of materials. More information on RILEM and its previous publications can be found on www.RILEM.net.

The RILEM State-of-the-Art Reports (STAR) are produced by the Technical Committees. They represent one of the most important outputs that RILEM generates – high level scientific and engineering reports that provide cutting edge knowledge in a given field. The work of the TCs is one of RILEM's key functions.

Members of a TC are experts in their field and give their time freely to share their expertise. As a result, the broader scientific community benefits greatly from RILEM's activities.

RILEM's stated objective is to disseminate this information as widely as possible to the scientific community. RILEM therefore considers the STAR reports of its TCs as of highest importance, and encourages their publication whenever possible.

The information in this and similar reports is mostly pre-normative in the sense that it provides the underlying scientific fundamentals on which standards and codes of practice are based. Without such a solid scientific basis, construction practice will be less than efficient or economical.

It is RILEM's hope that this information will be of wide use to the scientific community.

Indexed in SCOPUS, Google Scholar and SpringerLink.

Hassan Baaj · Orazio Baglieri
Editors

Crack-Healing of Asphalt Pavement Materials

State-of-the-Art Report of the RILEM Technical Committee 278-CHA

Springer

Editors
Hassan Baaj
Department of Civil and Environmental Engineering
University of Waterloo
Waterloo, ON, Canada

Orazio Baglieri
Department of Environmental, Land and Infrastructure Engineering
Politecnico di Torino
Turin, Italy

ISSN 2213-204X ISSN 2213-2031 (electronic)
RILEM State-of-the-Art Reports
ISBN 978-3-031-99442-5 ISBN 978-3-031-99443-2 (eBook)
https://doi.org/10.1007/978-3-031-99443-2

© RILEM 2025

No part of this work may be reproduced, stored in a retrieval system, or transmitted in any form or by any means, electronic, mechanical, photocopying, microfilming, recording or otherwise, without written permission from the Publisher, with the exception of any material supplied specifically for the purpose of being entered and executed on a computer system, for exclusive use by the purchaser of the work. Permission for use must always be obtained from the owner of the copyright: RILEM
The use of general descriptive names, registered names, trademarks, service marks, etc. in this publication does not imply, even in the absence of a specific statement, that such names are exempt from the relevant protective laws and regulations and therefore free for general use.
The publisher, the authors and the editors are safe to assume that the advice and information in this book are believed to be true and accurate at the date of publication. Neither the publisher nor the authors or the editors give a warranty, expressed or implied, with respect to the material contained herein or for any errors or omissions that may have been made. The publisher remains neutral with regard to jurisdictional claims in published maps and institutional affiliations.

This Springer imprint is published by the registered company Springer Nature Switzerland AG
The registered company address is: Gewerbestrasse 11, 6330 Cham, Switzerland

If disposing of this product, please recycle the paper.

TC 278-CHA Committee Members

Hassan Baaj, Chair
Orazio Baglieri, Deputy Chair
Ferhat Hammoum, TG Leader
Greet Leegwater, TG Leader
Salvatore Mangiafico, TG Leader
Amir Tabaković, TG Leader

Haya Almutairi, Roberto M. Aurilio, Hussain U. Bahia, Moises Bueno, Augusto Cannone Falchetto, Jean-Claude Carret, Alan Carter, Eshan V. Dave, Hervé Di Benedetto, Alvaro Garcia Hernandez, Bernhard Hofko, Yasmina Mahmoudi, Fabrizio Miglietta, José Norambuena-Contreras, Jorge Carvalho Pais, Christophe Petit, Laurent Porot, Maria Del Carmen Rubio-Gamez, Cédric Sauzeat, Pejoohan Tavassoti, Lucia Tsantilis, Michael P. Wistuba, Dandi Zhao

RILEM Publications

The following list is presents the global offer of RILEM Publications, sorted by series. Each publication is available in printed version and/or in online version.

RILEM Proceedings (PRO)

PRO 1: Durability of High Performance Concrete (ISBN: 2-912143-03-9; e-ISBN: 2-351580-12-5; e-ISBN: 2351580125); *Ed. H. Sommer*

PRO 2: Chloride Penetration into Concrete (ISBN: 2-912143-00-04; e-ISBN: 2912143454); *Eds. L.-O. Nilsson and J.-P. Ollivier*

PRO 3: Evaluation and Strengthening of Existing Masonry Structures (ISBN: 2-912143-02-0; e-ISBN: 2351580141); *Eds. L. Binda and C. Modena*

PRO 4: Concrete: From Material to Structure (ISBN: 2-912143-04-7; e-ISBN: 2351580206); *Eds. J.-P. Bournazel and Y. Malier*

PRO 5: The Role of Admixtures in High Performance Concrete (ISBN: 2-912143-05-5; e-ISBN: 2351580214); *Eds. J. G. Cabrera and R. Rivera-Villarreal*

PRO 6: High Performance Fiber Reinforced Cement Composites—HPFRCC 3 (ISBN: 2-912143-06-3; e-ISBN: 2351580222); *Eds. H. W. Reinhardt and A. E. Naaman*

PRO 7: 1st International RILEM Symposium on Self-Compacting Concrete (ISBN: 2-912143-09-8; e-ISBN: 2912143721); *Eds. Å. Skarendahl and Ö. Petersson*

PRO 8: International RILEM Symposium on Timber Engineering (ISBN: 2-912143-10-1; e-ISBN: 2351580230); *Ed. L. Boström*

PRO 9: 2nd International RILEM Symposium on Adhesion between Polymers and Concrete ISAP '99 (ISBN: 2-912143-11-X; e-ISBN: 2351580249); *Eds. Y. Ohama and M. Puterman*

PRO 10: 3rd International RILEM Symposium on Durability of Building and Construction Sealants (ISBN: 2-912143-13-6; e-ISBN: 2351580257); *Ed. A. T. Wolf*

PRO 11: 4th International RILEM Conference on Reflective Cracking in Pavements (ISBN: 2-912143-14-4; e-ISBN: 2351580265); *Eds. A. O. Abd El Halim, D. A. Taylor and El H. H. Mohamed*

PRO 12: International RILEM Workshop on Historic Mortars: Characteristics and Tests (ISBN: 2-912143-15-2; e-ISBN: 2351580273); *Eds. P. Bartos, C. Groot and J. J. Hughes*

PRO 13: 2nd International RILEM Symposium on Hydration and Setting (ISBN: 2-912143-16-0; e-ISBN: 2351580281); *Ed. A. Nonat*

PRO 14: Integrated Life-Cycle Design of Materials and Structures—ILCDES 2000 (ISBN: 951-758-408-3; e-ISBN: 235158029X); (ISSN: 0356-9403); *Ed. S. Sarja*

PRO 15: Fifth RILEM Symposium on Fibre-Reinforced Concretes (FRC)—BEFIB'2000 (ISBN: 2-912143-18-7; e-ISBN: 291214373X); *Eds. P. Rossi and G. Chanvillard*

PRO 16: Life Prediction and Management of Concrete Structures (ISBN: 2-912143-19-5; e-ISBN: 2351580303); *Ed. D. Naus*

PRO 17: Shrinkage of Concrete—Shrinkage 2000 (ISBN: 2-912143-20-9; e-ISBN: 2351580311); *Eds. V. Baroghel-Bouny and P.-C. Aïtcin*

PRO 18: Measurement and Interpretation of the On-Site Corrosion Rate (ISBN: 2-912143-21-7; e-ISBN: 235158032X); *Eds. C. Andrade, C. Alonso, J. Fullea, J. Polimon and J. Rodriguez*

PRO 19: Testing and Modelling the Chloride Ingress into Concrete (ISBN: 2-912143-22-5; e-ISBN: 2351580338); *Eds. C. Andrade and J. Kropp*

PRO 20: 1st International RILEM Workshop on Microbial Impacts on Building Materials (CD 02) (e-ISBN 978-2-35158-013-4); *Ed. M. Ribas Silva*

PRO 21: International RILEM Symposium on Connections between Steel and Concrete (ISBN: 2-912143-25-X; e-ISBN: 2351580346); *Ed. R. Eligehausen*

PRO 22: International RILEM Symposium on Joints in Timber Structures (ISBN: 2-912143-28-4; e-ISBN: 2351580354); *Eds. S. Aicher and H.-W. Reinhardt*

PRO 23: International RILEM Conference on Early Age Cracking in Cementitious Systems (ISBN: 2-912143-29-2; e-ISBN: 2351580362); *Eds. K. Kovler and A. Bentur*

PRO 24: 2nd International RILEM Workshop on Frost Resistance of Concrete (ISBN: 2-912143-30-6; e-ISBN: 2351580370); *Eds. M. J. Setzer, R. Auberg and H.-J. Keck*

PRO 25: International RILEM Workshop on Frost Damage in Concrete (ISBN: 2-912143-31-4; e-ISBN: 2351580389); *Eds. D. J. Janssen, M. J. Setzer and M. B. Snyder*

PRO 26: International RILEM Workshop on On-Site Control and Evaluation of Masonry Structures (ISBN: 2-912143-34-9; e-ISBN: 2351580141); *Eds. L. Binda and R. C. de Vekey*

PRO 27: International RILEM Symposium on Building Joint Sealants (CD03; e-ISBN: 235158015X); *Ed. A. T. Wolf*

PRO 28: 6th International RILEM Symposium on Performance Testing and Evaluation of Bituminous Materials—PTEBM'03 (ISBN: 2-912143-35-7; e-ISBN: 978-2-912143-77-8); *Ed. M. N. Partl*

PRO 29: 2nd International RILEM Workshop on Life Prediction and Ageing Management of Concrete Structures (ISBN: 2-912143-36-5; e-ISBN: 2912143780); *Ed. D. J. Naus*

PRO 30: 4th International RILEM Workshop on High Performance Fiber Reinforced Cement Composites—HPFRCC 4 (ISBN: 2-912143-37-3; e-ISBN: 2912143799); *Eds. A. E. Naaman and H. W. Reinhardt*

PRO 31: International RILEM Workshop on Test and Design Methods for Steel Fibre Reinforced Concrete: Background and Experiences (ISBN: 2-912143-38-1; e-ISBN: 2351580168); *Eds. B. Schnütgen and L. Vandewalle*

PRO 32: International Conference on Advances in Concrete and Structures 2 vol. (ISBN (set): 2-912143-41-1; e-ISBN: 2351580176); *Eds. Ying-shu Yuan, Surendra P. Shah and Heng-lin Lü*

PRO 33: 3rd International Symposium on Self-Compacting Concrete (ISBN: 2-912143-42-X; e-ISBN: 2912143713); *Eds. Ó. Wallevik and I. Níelsson*

PRO 34: International RILEM Conference on Microbial Impact on Building Materials (ISBN: 2-912143-43-8; e-ISBN: 2351580184); *Ed. M. Ribas Silva*

PRO 35: International RILEM TC 186-ISA on Internal Sulfate Attack and Delayed Ettringite Formation (ISBN: 2-912143-44-6; e-ISBN: 2912143802); *Eds. K. Scrivener and J. Skalny*

PRO 36: International RILEM Symposium on Concrete Science and Engineering—A Tribute to Arnon Bentur (ISBN: 2-912143-46-2; e-ISBN: 2912143586); *Eds. K. Kovler, J. Marchand, S. Mindess and J. Weiss*

PRO 37: 5th International RILEM Conference on Cracking in Pavements—Mitigation, Risk Assessment and Prevention (ISBN: 2-912143-47-0; e-ISBN: 2912143764); *Eds. C. Petit, I. Al-Qadi and A. Millien*

PRO 38: 3rd International RILEM Workshop on Testing and Modelling the Chloride Ingress into Concrete (ISBN: 2-912143-48-9; e-ISBN: 2912143578); *Eds. C. Andrade and J. Kropp*

PRO 39: 6th International RILEM Symposium on Fibre-Reinforced Concretes—BEFIB 2004 (ISBN: 2-912143-51-9; e-ISBN: 2912143748); *Eds. M. Di Prisco, R. Felicetti and G. A. Plizzari*

PRO 40: International RILEM Conference on the Use of Recycled Materials in Buildings and Structures (ISBN: 2-912143-52-7; e-ISBN: 2912143756); *Eds. E. Vázquez, Ch. F. Hendriks and G. M. T. Janssen*

PRO 41: RILEM International Symposium on Environment-Conscious Materials and Systems for Sustainable Development (ISBN: 2-912143-55-1; e-ISBN: 2912143640); *Eds. N. Kashino and Y. Ohama*

PRO 42: SCC'2005—China: 1st International Symposium on Design, Performance and Use of Self-Consolidating Concrete (ISBN: 2-912143-61-6; e-ISBN: 2912143624); *Eds. Zhiwu Yu, Caijun Shi, Kamal Henri Khayat and Youjun Xie*

PRO 43: International RILEM Workshop on Bonded Concrete Overlays (e-ISBN: 2-912143-83-7); *Eds. J. L. Granju and J. Silfwerbrand*

PRO 44: 2nd International RILEM Workshop on Microbial Impacts on Building Materials (CD11) (e-ISBN: 2-912143-84-5); *Ed. M. Ribas Silva*

PRO 45: 2nd International Symposium on Nanotechnology in Construction, Bilbao (ISBN: 2-912143-87-X; e-ISBN: 2912143888); *Eds. Peter J. M. Bartos, Yolanda de Miguel and Antonio Porro*

PRO 46: Concrete Life'06—International RILEM-JCI Seminar on Concrete Durability and Service Life Planning: Curing, Crack Control, Performance in Harsh Environments (ISBN: 2-912143-89-6; e-ISBN: 291214390X); *Ed. K. Kovler*

PRO 47: International RILEM Workshop on Performance Based Evaluation and Indicators for Concrete Durability (ISBN: 978-2-912143-95-2; e-ISBN: 9782912143969); *Eds. V. Baroghel-Bouny, C. Andrade, R. Torrent and K. Scrivener*

PRO 48: 1st International RILEM Symposium on Advances in Concrete through Science and Engineering (e-ISBN: 2-912143-92-6); *Eds. J. Weiss, K. Kovler, J. Marchand, and S. Mindess*

PRO 49: International RILEM Workshop on High Performance Fiber Reinforced Cementitious Composites in Structural Applications (ISBN: 2-912143-93-4; e-ISBN: 2912143942); *Eds. G. Fischer and V. C. Li*

PRO 50: 1st International RILEM Symposium on Textile Reinforced Concrete (ISBN: 2-912143-97-7; e-ISBN: 2351580087); *Eds. Josef Hegger, Wolfgang Brameshuber and Norbert Will*

PRO 51: 2nd International Symposium on Advances in Concrete through Science and Engineering (ISBN: 2-35158-003-6; e-ISBN: 2-35158-002-8); *Eds. J. Marchand, B. Bissonnette, R. Gagné, M. Jolin and F. Paradis*

PRO 52: Volume Changes of Hardening Concrete: Testing and Mitigation (ISBN: 2-35158-004-4; e-ISBN: 2-35158-005-2); *Eds. O. M. Jensen, P. Lura and K. Kovler*

PRO 53: High Performance Fiber Reinforced Cement Composites—HPFRCC5 (ISBN: 978-2-35158-046-2; e-ISBN: 978-2-35158-089-9); *Eds. H. W. Reinhardt and A. E. Naaman*

PRO 54: 5th International RILEM Symposium on Self-Compacting Concrete (ISBN: 978-2-35158-047-9; e-ISBN: 978-2-35158-088-2); *Eds. G. De Schutter and V. Boel*

PRO 55: International RILEM Symposium Photocatalysis, Environment and Construction Materials (ISBN: 978-2-35158-056-1; e-ISBN: 978-2-35158-057-8); *Eds. P. Baglioni and L. Cassar*

PRO 56: International RILEM Workshop on Integral Service Life Modelling of Concrete Structures (ISBN 978-2-35158-058-5; e-ISBN: 978-2-35158-090-5); *Eds. R. M. Ferreira, J. Gulikers and C. Andrade*

PRO 57: RILEM Workshop on Performance of cement-based materials in aggressive aqueous environments (e-ISBN: 978-2-35158-059-2); *Ed. N. De Belie*

PRO 58: International RILEM Symposium on Concrete Modelling—CONMOD'08 (ISBN: 978-2-35158-060-8; e-ISBN: 978-2-35158-076-9); *Eds. E. Schlangen and G. De Schutter*

PRO 59: International RILEM Conference on On Site Assessment of Concrete, Masonry and Timber Structures—SACoMaTiS 2008 (ISBN set: 978-2-35158-061-5; e-ISBN: 978-2-35158-075-2); *Eds. L. Binda, M. di Prisco and R. Felicetti*

PRO 60: Seventh RILEM International Symposium on Fibre Reinforced Concrete: Design and Applications—BEFIB 2008 (ISBN: 978-2-35158-064-6; e-ISBN: 978-2-35158-086-8); *Ed. R. Gettu*

PRO 61: 1st International Conference on Microstructure Related Durability of Cementitious Composites 2 vol., (ISBN: 978-2-35158-065-3; e-ISBN: 978-2-35158-084-4); *Eds. W. Sun, K. van Breugel, C. Miao, G. Ye and H. Chen*

PRO 62: NSF/RILEM Workshop: In-situ Evaluation of Historic Wood and Masonry Structures (e-ISBN: 978-2-35158-068-4); *Eds. B. Kasal, R. Anthony and M. Drdácký*

PRO 63: Concrete in Aggressive Aqueous Environments: Performance, Testing and Modelling, 2 vol., (ISBN: 978-2-35158-071-4; e-ISBN: 978-2-35158-082-0); *Eds. M. G. Alexander and A. Bertron*

PRO 64: Long Term Performance of Cementitious Barriers and Reinforced Concrete in Nuclear Power Plants and Waste Management—NUCPERF 2009 (ISBN: 978-2-35158-072-1; e-ISBN: 978-2-35158-087-5); *Eds. V. L'Hostis, R. Gens and C. Gallé*

PRO 65: Design Performance and Use of Self-consolidating Concrete—SCC'2009 (ISBN: 978-2-35158-073-8; e-ISBN: 978-2-35158-093-6); *Eds. C. Shi, Z. Yu, K. H. Khayat and P. Yan*

PRO 66: 2nd International RILEM Workshop on Concrete Durability and Service Life Planning—ConcreteLife'09 (ISBN: 978-2-35158-074-5; ISBN: 978-2-35158-074-5); *Ed. K. Kovler*

PRO 67: Repairs Mortars for Historic Masonry (e-ISBN: 978-2-35158-083-7); *Ed. C. Groot*

PRO 68: Proceedings of the 3rd International RILEM Symposium on 'Rheology of Cement Suspensions such as Fresh Concrete' (ISBN 978-2-35158-091-2; e-ISBN: 978-2-35158-092-9); *Eds. O. H. Wallevik, S. Kubens and S. Oesterheld*

PRO 69: 3rd International PhD Student Workshop on 'Modelling the Durability of Reinforced Concrete' (ISBN: 978-2-35158-095-0); *Eds. R. M. Ferreira, J. Gulikers and C. Andrade*

PRO 70: 2nd International Conference on 'Service Life Design for Infrastructure' (ISBN set: 978-2-35158-096-7, e-ISBN: 978-2-35158-097-4); *Eds. K. van Breugel, G. Ye and Y. Yuan*

PRO 71: Advances in Civil Engineering Materials—'The 50-year Teaching Anniversary of Prof. Sun Wei' (ISBN: 978-2-35158-098-1; e-ISBN: 978-2-35158-099-8); *Eds. C. Miao, G. Ye and H. Chen*

PRO 72: First International Conference on 'Advances in Chemically-Activated Materials—CAM'2010' (2010), 264 pp., ISBN: 978-2-35158-101-8; e-ISBN: 978-2-35158-115-5; *Eds. Caijun Shi and Xiaodong Shen*

PRO 73: 2nd International Conference on 'Waste Engineering and Management—ICWEM 2010' (2010), 894 pp., ISBN: 978-2-35158-102-5; e-ISBN: 978-2-35158-103-2, *Eds. J. Zh. Xiao, Y. Zhang, M. S. Cheung and R. Chu*

PRO 74: International RILEM Conference on 'Use of Superabsorbent Polymers and Other New Additives in Concrete' (2010) 374 pp., ISBN: 978-2-35158-104-9; e-ISBN: 978-2-35158-105-6; *Eds. O. M. Jensen, M. T. Hasholt, and S. Laustsen*

PRO 75: International Conference on 'Material Science—2nd ICTRC—Textile Reinforced Concrete—Theme 1' (2010) 436 pp., ISBN: 978-2-35158-106-3; e-ISBN: 978-2-35158-107-0; *Ed. W. Brameshuber*

PRO 76: International Conference on 'Material Science—HetMat—Modelling of Heterogeneous Materials—Theme 2' (2010) 255 pp., ISBN: 978-2-35158-108-7; e-ISBN: 978-2-35158-109-4; *Ed. W. Brameshuber*

PRO 77: International Conference on 'Material Science—AdIPoC—Additions Improving Properties of Concrete—Theme 3' (2010) 459 pp., ISBN: 978-2-35158-110-0; e-ISBN: 978-2-35158-111-7; *Ed. W. Brameshuber*

PRO 78: 2nd Historic Mortars Conference and RILEM TC 203-RHM Final Workshop—HMC2010 (2010) 1416 pp., e-ISBN: 978-2-35158-112-4; *Eds. J. Válek, C. Groot and J. J. Hughes*

PRO 79: International RILEM Conference on Advances in Construction Materials Through Science and Engineering (2011) 213 pp., ISBN: 978-2-35158-116-2, e-ISBN: 978-2-35158-117-9; *Eds. Christopher Leung and K. T. Wan*

PRO 80: 2nd International RILEM Conference on Concrete Spalling due to Fire Exposure (2011) 453 pp., ISBN: 978-2-35158-118-6; e-ISBN: 978-2-35158-119-3; *Eds. E. A. B. Koenders and F. Dehn*

PRO 81: 2nd International RILEM Conference on Strain Hardening Cementitious Composites (SHCC2-Rio) (2011) 451 pp., ISBN: 978-2-35158-120-9; e-ISBN: 978-2-35158-121-6; *Eds. R. D. Toledo Filho, F. A. Silva, E. A. B. Koenders and E. M. R. Fairbairn*

PRO 82: 2nd International RILEM Conference on Progress of Recycling in the Built Environment (2011) 507 pp., e-ISBN: 978-2-35158-122-3; *Eds. V. M. John, E. Vazquez, S. C. Angulo and C. Ulsen*

PRO 83: 2nd International Conference on Microstructural-related Durability of Cementitious Composites (2012) 250 pp., ISBN: 978-2-35158-129-2; e-ISBN: 978-2-35158-123-0; *Eds. G. Ye, K. van Breugel, W. Sun and C. Miao*

PRO 84: CONSEC13—Seventh International Conference on Concrete under Severe Conditions—Environment and Loading (2013) 1930 pp., ISBN: 978-2-35158-124-7; e-ISBN: 978-2-35158-134-6; *Eds. Z. J. Li, W. Sun, C. W. Miao, K. Sakai, O. E. Gjorv and N. Banthia*

PRO 85: RILEM-JCI International Workshop on Crack Control of Mass Concrete and Related issues concerning Early-Age of Concrete Structures—ConCrack 3—Control of Cracking in Concrete Structures 3 (2012) 237 pp., ISBN: 978-2-35158-125-4; e-ISBN: 978-2-35158-126-1; *Eds. F. Toutlemonde and J.-M. Torrenti*

PRO 86: International Symposium on Life Cycle Assessment and Construction (2012) 414 pp., ISBN: 978-2-35158-127-8, e-ISBN: 978-2-35158-128-5; *Eds. A. Ventura and C. de la Roche*

PRO 87: UHPFRC 2013—RILEM-fib-AFGC International Symposium on Ultra-High Performance Fibre-Reinforced Concrete (2013), ISBN: 978-2-35158-130-8, e-ISBN: 978-2-35158-131-5; *Eds. F. Toutlemonde*

PRO 88: 8th RILEM International Symposium on Fibre Reinforced Concrete (2012) 344 pp., ISBN: 978-2-35158-132-2; e-ISBN: 978-2-35158-133-9; *Eds. Joaquim A. O. Barros*

PRO 89: RILEM International workshop on performance-based specification and control of concrete durability (2014) 678 pp., ISBN: 978-2-35158-135-3; e-ISBN: 978-2-35158-136-0; *Eds. D. Bjegović, H. Beushausen and M. Serdar*

PRO 90: 7th RILEM International Conference on Self-Compacting Concrete and of the 1st RILEM International Conference on Rheology and Processing of Construction Materials (2013) 396 pp., ISBN: 978-2-35158-137-7; e-ISBN: 978-2-35158-138-4; *Eds. Nicolas Roussel and Hela Bessaies-Bey*

PRO 91: CONMOD 2014—RILEM International Symposium on Concrete Modelling (2014), ISBN: 978-2-35158-139-1; e-ISBN: 978-2-35158-140-7; *Eds. Kefei Li, Peiyu Yan and Rongwei Yang*

PRO 92: CAM 2014—2nd International Conference on advances in chemically-activated materials (2014) 392 pp., ISBN: 978-2-35158-141-4; e-ISBN: 978-2-35158-142-1; *Eds. Caijun Shi and Xiadong Shen*

PRO 93: SCC 2014—3rd International Symposium on Design, Performance and Use of Self-Consolidating Concrete (2014) 438 pp., ISBN: 978-2-35158-143-8; e-ISBN: 978-2-35158-144-5; *Eds. Caijun Shi, Zhihua Ou and Kamal H. Khayat*

PRO 94 (online version): HPFRCC-7—7th RILEM conference on High performance fiber reinforced cement composites (2015), e-ISBN: 978-2-35158-146-9; *Eds. H. W. Reinhardt, G. J. Parra-Montesinos and H. Garrecht*

PRO 95: International RILEM Conference on Application of superabsorbent polymers and other new admixtures in concrete construction (2014), ISBN: 978-2-35158-147-6; e-ISBN: 978-2-35158-148-3; *Eds. Viktor Mechtcherine and Christof Schroefl*

PRO 96 (online version): XIII DBMC: XIII International Conference on Durability of Building Materials and Components (2015), e-ISBN: 978-2-35158-149-0; *Eds. M. Quattrone and V. M. John*

PRO 97: SHCC3—3rd International RILEM Conference on Strain Hardening Cementitious Composites (2014), ISBN: 978-2-35158-150-6; e-ISBN: 978-2-35158-151-3; *Eds. E. Schlangen, M. G. Sierra Beltran, M. Lukovic and G. Ye*

PRO 98: FERRO-11—11th International Symposium on Ferrocement and 3rd ICTRC—International Conference on Textile Reinforced Concrete (2015), ISBN: 978-2-35158-152-0; e-ISBN: 978-2-35158-153-7; *Ed. W. Brameshuber*

PRO 99 (online version): ICBBM 2015—1st International Conference on Bio-Based Building Materials (2015), e-ISBN: 978-2-35158-154-4; *Eds. S. Amziane and M. Sonebi*

PRO 100: SCC16—RILEM Self-Consolidating Concrete Conference (2016), ISBN: 978-2-35158-156-8; e-ISBN: 978-2-35158-157-5; *Ed. Kamal H. Kayat*

PRO 101 (online version): III Progress of Recycling in the Built Environment (2015), e-ISBN: 978-2-35158-158-2; *Eds I. Martins, C. Ulsen and S. C. Angulo*

PRO 102 (online version): RILEM Conference on Microorganisms-Cementitious Materials Interactions (2016), e-ISBN: 978-2-35158-160-5; *Eds. Alexandra Bertron, Henk Jonkers and Virginie Wiktor*

PRO 103 (online version): ACESC'16—Advances in Civil Engineering and Sustainable Construction (2016), e-ISBN: 978-2-35158-161-2; *Eds. T.Ch. Madhavi, G. Prabhakar, Santhosh Ram and P. M. Rameshwaran*

PRO 104 (online version): SSCS'2015—Numerical Modeling—Strategies for Sustainable Concrete Structures (2015), e-ISBN: 978-2-35158-162-9

PRO 105: 1st International Conference on UHPC Materials and Structures (2016), ISBN: 978-2-35158-164-3; e-ISBN: 978-2-35158-165-0

PRO 106: AFGC-ACI-fib-RILEM International Conference on Ultra-High-Performance Fibre-Reinforced Concrete—UHPFRC 2017 (2017), ISBN: 978-2-35158-166-7; e-ISBN: 978-2-35158-167-4; *Eds. François Toutlemonde and Jacques Resplendino*

PRO 107 (online version): XIV DBMC—14th International Conference on Durability of Building Materials and Components (2017), e-ISBN: 978-2-35158-159-9; *Eds. Geert De Schutter, Nele De Belie, Arnold Janssens and Nathan Van Den Bossche*

PRO 108: MSSCE 2016—Innovation of Teaching in Materials and Structures (2016), ISBN: 978-2-35158-178-0; e-ISBN: 978-2-35158-179-7; *Ed. Per Goltermann*

PRO 109 (2 volumes): MSSCE 2016—Service Life of Cement-Based Materials and Structures (2016), ISBN Vol. 1: 978-2-35158-170-4; Vol. 2: 978-2-35158-171-4; Set Vol. 1&2: 978-2-35158-172-8; e-ISBN: 978-2-35158-173-5; *Eds. Miguel Azenha, Ivan Gabrijel, Dirk Schlicke, Terje Kanstad and Ole Mejlhede Jensen*

PRO 110: MSSCE 2016—Historical Masonry (2016), ISBN: 978-2-35158-178-0; e-ISBN: 978-2-35158-179-7; *Eds. Inge Rörig-Dalgaard and Ioannis Ioannou*

PRO 111: MSSCE 2016—Electrochemistry in Civil Engineering (2016); ISBN: 978-2-35158-176-6; e-ISBN: 978-2-35158-177-3; *Ed. Lisbeth M. Ottosen*

PRO 112: MSSCE 2016—Moisture in Materials and Structures (2016), ISBN: 978-2-35158-178-0; e-ISBN: 978-2-35158-179-7; *Eds. Kurt Kielsgaard Hansen, Carsten Rode and Lars-Olof Nilsson*

PRO 113: MSSCE 2016—Concrete with Supplementary Cementitious Materials (2016), ISBN: 978-2-35158-178-0; e-ISBN: 978-2-35158-179-7; *Eds. Ole Mejlhede Jensen, Konstantin Kovler and Nele De Belie*

PRO 114: MSSCE 2016—Frost Action in Concrete (2016), ISBN: 978-2-35158-182-7; e-ISBN: 978-2-35158-183-4; *Eds. Marianne Tange Hasholt, Katja Fridh and R. Doug Hooton*

PRO 115: MSSCE 2016—Fresh Concrete (2016), ISBN: 978-2-35158-184-1; e-ISBN: 978-2-35158-185-8; *Eds. Lars N. Thrane, Claus Pade, Oldrich Svec and Nicolas Roussel*

PRO 116: BEFIB 2016—9th RILEM International Symposium on Fiber Reinforced Concrete (2016), ISBN: 978-2-35158-187-2; e-ISBN: 978-2-35158-186-5; *Eds. N. Banthia, M. di Prisco and S. Soleimani-Dashtaki*

PRO 117: 3rd International RILEM Conference on Microstructure Related Durability of Cementitious Composites (2016), ISBN: 978-2-35158-188-9; e-ISBN: 978-2-35158-189-6; *Eds. Changwen Miao, Wei Sun, Jiaping Liu, Huisu Chen, Guang Ye and Klaas van Breugel*

PRO 118 (4 volumes): International Conference on Advances in Construction Materials and Systems (2017), ISBN Set: 978-2-35158-190-2; Vol. 1: 978-2-35158-193-3; Vol. 2: 978-2-35158-194-0; Vol. 3: ISBN: 978-2-35158-195-7; Vol. 4: ISBN: 978-2-35158-196-4; e-ISBN: 978-2-35158-191-9; *Ed. Manu Santhanam*

PRO 119 (online version): ICBBM 2017—Second International RILEM Conference on Bio-based Building Materials, (2017), e-ISBN: 978-2-35158-192-6; *Ed. Sofiane Amziane*

PRO 120 (2 volumes): EAC-02—2nd International RILEM/COST Conference on Early Age Cracking and Serviceability in Cement-based Materials and Structures, (2017), Vol. 1: 978-2-35158-199-5, Vol. 2: 978-2-35158-200-8, Set: 978-2-35158-197-1, e-ISBN: 978-2-35158-198-8; *Eds. Stéphanie Staquet and Dimitrios Aggelis*

PRO 121 (2 volumes): SynerCrete18: Interdisciplinary Approaches for Cement-based Materials and Structural Concrete: Synergizing Expertise and Bridging Scales of Space and Time, (2018), Set: 978-2-35158-202-2, Vol.1: 978-2-35158-211-4, Vol.2: 978-2-35158-212-1, e-ISBN: 978-2-35158-203-9; *Eds. Miguel Azenha, Dirk Schlicke, Farid Benboudjema, Agnieszka Knoppik*

PRO 122: SCC'2018 China—Fourth International Symposium on Design, Performance and Use of Self-Consolidating Concrete, (2018), ISBN: 978-2-35158-204-6, e-ISBN: 978-2-35158-205-3; *Eds. C. Shi, Z. Zhang, K. H. Khayat*

PRO 123: Final Conference of RILEM TC 253-MCI: Microorganisms-Cementitious Materials Interactions (2018), Set: 978-2-35158-207-7, Vol. 1: 978-2-35158-209-1, Vol. 2: 978-2-35158-210-7, e-ISBN: 978-2-35158-206-0; *Ed. Alexandra Bertron*

PRO 124 (online version): Fourth International Conference Progress of Recycling in the Built Environment (2018), e-ISBN: 978-2-35158-208-4; *Eds. Isabel M. Martins, Carina Ulsen, Yury Villagran*

PRO 125 (online version): SLD4—4th International Conference on Service Life Design for Infrastructures (2018), e-ISBN: 978-2-35158-213-8; *Eds. Guang Ye, Yong Yuan, Claudia Romero Rodriguez, Hongzhi Zhang, Branko Savija*

PRO 126: Workshop on Concrete Modelling and Material Behaviour in honor of Professor Klaas van Breugel (2018), ISBN: 978-2-35158-214-5, e-ISBN: 978-2-35158-215-2; *Ed. Guang Ye*

PRO 127 (online version): CONMOD2018—Symposium on Concrete Modelling (2018), e-ISBN: 978-2-35158-216-9; *Eds. Erik Schlangen, Geert de Schutter, Branko Savija, Hongzhi Zhang, Claudia Romero Rodriguez*

PRO 128: SMSS2019—International Conference on Sustainable Materials, Systems and Structures (2019), ISBN: 978-2-35158-217-6, e-ISBN: 978-2-35158-218-3

PRO 129: 2nd International Conference on UHPC Materials and Structures (UHPC2018-China), ISBN: 978-2-35158-219-0, e-ISBN: 978-2-35158-220-6

PRO 130: 5th Historic Mortars Conference (2019), ISBN: 978-2-35158-221-3, e-ISBN: 978-2-35158-222-0; *Eds. José Ignacio Álvarez, José María Fernández, Íñigo Navarro, Adrián Durán, Rafael Sirera*

PRO 131 (online version): 3rd International Conference on Bio-Based Building Materials (ICBBM2019), e-ISBN: 978-2-35158-229-9; *Eds. Mohammed Sonebi, Sofiane Amziane, Jonathan Page*

PRO 132: IRWRMC'18—International RILEM Workshop on Rheological Measurements of Cement-based Materials (2018), ISBN: 978-2-35158-230-5, e-ISBN: 978-2-35158-231-2; *Eds. Chafika Djelal, Yannick Vanhove*

PRO 133 (online version): CO2STO2019—International Workshop CO2 Storage in Concrete (2019), e-ISBN: 978-2-35158-232-9; *Eds. Assia Djerbi, Othman Omikrine-Metalssi, Teddy Fen-Chong*

PRO 134: 3rd ACF/HNU International Conference on UHPC Materials and Structures—UHPC'2020, ISBN: 978-2-35158-233-6, e-ISBN: 978-2-35158-234-3; *Eds. Caijun Shi and Jiaping Liu*

PRO 135: Fourth International Conference on Chemically Activated Materials (CAM2021), ISBN: 978-2-35158-235-0, e-ISBN: 978-2-35158-236-7; *Eds. Caijun Shi and Xiang Hu*

PRO 136: Lime Based Materials for Repairing Historic Structures (2022), ISBN: 978-2-35158-240-4, e-ISBN: 978-2-35158 237-4; *Eds. Ioanna Papayianni and Jan Válek*

PRO 137 (online version): V International Conference Progress of Recycling in the Built Environment (2023), e-ISBN: 978-2-35158-238-1; *Eds. Ines Döring, Yury Villagrán Zaccardi, Anette Müller*

RILEM Reports (REP)

Report 19: Considerations for Use in Managing the Aging of Nuclear Power Plant Concrete Structures (ISBN: 2-912143-07-1); *Ed. D. J. Naus*

Report 20: Engineering and Transport Properties of the Interfacial Transition Zone in Cementitious Composites (ISBN: 2-912143-08-X); *Eds. M. G. Alexander, G. Arliguie, G. Ballivy, A. Bentur and J. Marchand*

Report 21: Durability of Building Sealants (ISBN: 2-912143-12-8); *Ed. A. T. Wolf*

Report 22: Sustainable Raw Materials—Construction and Demolition Waste (ISBN: 2-912143-17-9); *Eds. C. F. Hendriks and H. S. Pietersen*

Report 23: Self-Compacting Concrete state-of-the-art report (ISBN: 2-912143-23-3); *Eds. Å. Skarendahl and Ö. Petersson*

Report 24: Workability and Rheology of Fresh Concrete: Compendium of Tests (ISBN: 2-912143-32-2); *Eds. P. J. M. Bartos, M. Sonebi and A. K. Tamimi*

Report 25: Early Age Cracking in Cementitious Systems (ISBN: 2-912143-33-0); *Ed. A. Bentur*

Report 26: Towards Sustainable Roofing (Joint Committee CIB/RILEM) (CD 07) (e-ISBN 978-2-912143-65-5); *Eds. Thomas W. Hutchinson and Keith Roberts*

Report 27: Condition Assessment of Roofs (Joint Committee CIB/RILEM) (CD 08) (e-ISBN 978-2-912143-66-2); *Ed. CIB W 83/RILEM TC166-RMS*

Report 28: Final report of RILEM TC 167-COM 'Characterisation of Old Mortars with Respect to Their Repair' (ISBN: 978-2-912143-56-3); *Eds. C. Groot, G. Ashall and J. Hughes*

Report 29: Pavement Performance Prediction and Evaluation (PPPE): Interlaboratory Tests (e-ISBN: 2-912143-68-3); *Eds. M. Partl and H. Piber*

Report 30: Final Report of RILEM TC 198-URM 'Use of Recycled Materials' (ISBN: 2-912143-82-9; e-ISBN: 2-912143-69-1); *Eds. Ch. F. Hendriks, G. M. T. Janssen and E. Vázquez*

Report 31: Final Report of RILEM TC 185-ATC 'Advanced testing of cement-based materials during setting and hardening' (ISBN: 2-912143-81-0; e-ISBN: 2-912143-70-5); *Eds. H. W. Reinhardt and C. U. Grosse*

Report 32: Probabilistic Assessment of Existing Structures. A JCSS publication (ISBN 2-912143-24-1); *Ed. D. Diamantidis*

Report 33: State-of-the-Art Report of RILEM Technical Committee TC 184-IFE 'Industrial Floors' (ISBN 2-35158-006-0); *Ed. P. Seidler*

Report 34: Report of RILEM Technical Committee TC 147-FMB 'Fracture mechanics applications to anchorage and bond' Tension of Reinforced Concrete

Prisms—Round Robin Analysis and Tests on Bond (e-ISBN 2-912143-91-8); *Eds. L. Elfgren and K. Noghabai*

Report 35: Final Report of RILEM Technical Committee TC 188-CSC 'Casting of Self Compacting Concrete' (ISBN 2-35158-001-X; e-ISBN: 2-912143-98-5); *Eds. Å. Skarendahl and P. Billberg*

Report 36: State-of-the-Art Report of RILEM Technical Committee TC 201-TRC 'Textile Reinforced Concrete' (ISBN 2-912143-99-3); *Ed. W. Brameshuber*

Report 37: State-of-the-Art Report of RILEM Technical Committee TC 192-ECM 'Environment-conscious construction materials and systems' (ISBN: 978-2-35158-053-0); *Eds. N. Kashino, D. Van Gemert and K. Imamoto*

Report 38: State-of-the-Art Report of RILEM Technical Committee TC 205-DSC 'Durability of Self-Compacting Concrete' (ISBN: 978-2-35158-048-6); *Eds. G. De Schutter and K. Audenaert*

Report 39: Final Report of RILEM Technical Committee TC 187-SOC 'Experimental determination of the stress-crack opening curve for concrete in tension' (ISBN 978-2-35158-049-3); *Ed. J. Planas*

Report 40: State-of-the-Art Report of RILEM Technical Committee TC 189-NEC 'Non-Destructive Evaluation of the Penetrability and Thickness of the Concrete Cover' (ISBN 978-2-35158-054-7); *Eds. R. Torrent and L. Fernández Luco*

Report 41: State-of-the-Art Report of RILEM Technical Committee TC 196-ICC 'Internal Curing of Concrete' (ISBN 978-2-35158-009-7); *Eds. K. Kovler and O. M. Jensen*

Report 42: 'Acoustic Emission and Related Non-destructive Evaluation Techniques for Crack Detection and Damage Evaluation in Concrete'—Final Report of RILEM Technical Committee 212-ACD (e-ISBN: 978-2-35158-100-1); *Ed. M. Ohtsu*

Report 45: Repair Mortars for Historic Masonry—State-of-the-Art Report of RILEM Technical Committee TC 203-RHM (e-ISBN: 978-2-35158-163-6); *Eds. Paul Maurenbrecher and Caspar Groot*

Report 46: Surface delamination of concrete industrial floors and other durability related aspects guide—Report of RILEM Technical Committee TC 268-SIF (e-ISBN: 978-2-35158-201-5); *Ed. Valerie Pollet*

Contents

Introduction

Hassan Baaj and Orazio Baglieri

The term "Self-healing materials" refers to materials that are capable of fully or partially repairing themselves when they experience damage with a local or global functionality loss. Local functionality loss occurs when a section of a material or structure exhibits degraded performance when compared with the rest of the material. Global functionality loss occurs when the material or structure exhibits degraded performance compared to its performance prior to any exposure to operational actions.

Self-healing materials can be categorized into two main groups: intrinsic and extrinsic [1]. Intrinsic self-healing is inherent to the material due to a dynamic cross-linking system that enables repair mechanisms without the need for external agents. Extrinsic self-healing is attributed to an added phase or action that is specifically used to activate healing phenomena. Due to their unique ability to restore damage, potential application of self-healing materials has been widely investigated in various industries, including aerospace, automotive, electronics, and construction [2, 3].

Fatigue damage is one of the main distress types affecting flexible pavement in road constructions. This distress is caused by repeated traffic loading and environmental actions that lead to a formation of diffused cracking. The presence of cracks on a road's surface undermines the structural integrity of the pavement and creates pathways for the intrusion of water into the lower layers and into the subgrade, with detrimental effects associated to moisture damage. Pavement cracking also results in functional failure of pavement with negative effects in terms of safety and riding comfort for drivers.

H. Baaj
Department of Civil and Environmental Engineering, University of Waterloo, 200 University Ave W, N2L 3G1, Waterloo, ON, Canada

O. Baglieri (✉)
Department of Environment, Land and Infrastructure Engineering, Politecnico di Torino, 24, c.so Duca degli Abruzzi, 10129 Turin, Italy
e-mail: orazio.baglieri@polito.it

© The Author(s), under exclusive license to Springer Nature Switzerland AG 2025
H. Baaj and O. Baglieri (eds.), *Crack-Healing of Asphalt Pavement Materials*, RILEM State-of-the-Art Reports 45, https://doi.org/10.1007/978-3-031-99443-2_1

Asphalt mixtures used in flexible pavements are characterized by intrinsic self-healing properties [4]. These mixtures repair themselves when a sufficient rest period elapses between external loadings. Due to intermittent nature of traffic, in which rest periods exist between each axle loading application, self-healing of micro-cracks contributes to prolonging the actual fatigue life of pavement. Therefore, the study of self-healing phenomena in asphalt materials represents a logical progression in understanding their fatigue and cracking behaviour in order to properly address the design of mixtures and improve pavement performance. This implies less road maintenance will be needed, leading to fewer traffic disruptions due to maintenance operations. This will also reduce the impact in terms of sustainability due to a reduction of natural resource consumption, reduction of CO_2 emissions, and a reduction of costs.

In order to further improve the self-repairing potential of asphalt pavements, many research efforts have been made in recent years to develop extrinsic self-healing technologies for asphalt mixtures [5]. Such technologies entail the incorporation of self-healing agents into the mixtures (nanoparticles, encapsulated rejuvenators, metallic conductive fibres). Depending on the approach adopted, healing phenomena are activated without any external action or may require the application of an external action, consisting in alternating electromagnetic field for induction and microwave heating.

Despite the fact that the self-healing of asphalt materials is a well-recognized phenomenon, its quantitative evaluation and modelling is difficult and still remains a challenging task. Due to the inherent complexity of healing mechanisms and the presence of many biasing effects, there is currently no standardized testing protocol to assess the self-healing properties for either asphalt binders or mixtures [6].

In the general context outlined above, the RILEM Technical Committee 278-CHA (Crack-healing of Asphalt Pavement Materials) aimed to provide a contribution in the advancement of knowledge in this specific area of research. The Technical Committee (TC) was active between 2016 and 2023, with the primary tasks of conducting literature review, laboratory experimentation through round-robin test, and modelling of laboratory test results.

This book presents the state-of-the-art report (STAR) based on the literature review work done by TC 278-CHA. The contents of the STAR cover several aspects of self-healing of asphalt pavements, as follows.

Chapter 2 deals with the mechanisms driving loss of performance and restoration of asphalt pavement materials. Relevance of intrinsic self-healing for asphalt pavement performance is discussed as well as the way of incorporating self-healing in the pavement design process. The chapter describes the main biasing effects that come into play in fatigue and healing evaluation and underlines the importance of tackling the study of healing phenomena at different scales. Definitions on crack-healing and restoration are finally proposed with the aim of standardizing terminology to be uniformly used in this research area.

Chapter 3 reviews extrinsic self-healing technologies currently available to aid asphalt's natural self-healing abilities. These technologies include nanoparticles, induction heating, microwave heating, rejuvenator encapsulation and hybrid (combination of induction and rejuvenator encapsulation). The existing technologies are presented and critically analyzed with respect to their functional design and performance. Future developments of self-healing asphalt technology are also outlined.

Chapter 4 provides an overview of the variety of experimental methods used to induce damage in the material and to allow the self-healing phenomena to manifest themselves when the material is left at rest, with an attempt at classifying various methods based on several relevant factors. Damage can be induced by means of both fatigue or fracture modes, requiring different testing protocols and theoretical approaches. When focusing on loading patterns and rest periods, testing methods can be generally classified as storage, intermittent and hybrid healing tests, depending on how rest phases are introduced in the overall testing sequence. The chapter also analyses various laboratory testing configuration, coupled with various analytical interpretations, to evaluate healing performance of asphalt materials at different scales (binder, mastic, mortar, and mixture). Several healing parameters introduced by researchers to measure the healing performance of asphalt materials are also presented.

Chapter 5 illustrates diverse models developed and proposed to analyze and estimate self-healing in asphalt materials. These models are based on experimental observations and numerical simulations, aiming to describe the healing response of materials under different thermal and loading conditions. The chapter highlights the existence of different modelling approaches that can be classified based on the scales of observation (nano-/micro-, intermediate-, and macro-). It also highlights the lack of a unified approach in current research.

Book Backmatter contains some final remarks and outlines future directions of research in the area of self-healing of asphalt pavement materials.

The chapters were designed to stand alone, allowing interested readers to select and read them independently without having to go through the entire book.

References

1. Pezzin, S. H. (2023). Mechanism of extrinsic and intrinsic self-healing in polymer systems. In N. Hameed, J. C. Capricho, N. Salim, & S. Thomas (Eds.), *Multifunctional epoxy resins. Self-healing thermally and electrically conductive resins* (pp. 107–138). Springer.
2. Bekas, D. G., Tsirka, K., Baltzis, D., & Paipetis, A. S. (2016). Self-healing materials: A review of advances in materials, evaluation, characterization and monitoring techniques. *Composites, Part B: Engineering, 87*, 92–119.
3. Sapana, J. (2014). Synthesis, mechanism, and application of self-healing materials. *Biomedical Materials and Devices, 2*, 225–240.
4. Little, D. N., Bhasin, A., & Darabi, M. K. (2015). Damage healing in asphalt pavements: Theory, mechanisms, measurement, and modeling. In *Advances in asphalt materials* (pp. 205–242). Woodhead Publishing.
5. Tabaković, A., & Schlangen, E. (2016). Self-healing technology for asphalt pavements. In M. D. Hager, S. van der Zwaag, & U. S. Schubert (Eds.), *Self-healing materials* (pp. 285–306). Springer International Publishing.

6. Leegwater, G., Taboković, A., Baglieri, O., Hammoum, F., & Baaj, H. (2022). Terms and definitions on crack-healing and restoration of mechanical properties in bituminous materials. In *Proceedings of the RILEM International Symposium on Bituminous Materials: ISBM Lyon 2020* (pp. 47–53).

Intrinsic Self-Healing of Asphalt Pavements

Greet Leegwater, Ferhat Hammoum, Amir Tabaković, Orazio Baglieri, and Hassan Baaj

Abstract Intrinsic self-healing refers to the inherent ability of asphalt materials to autonomously repair micro-cracks and damage, without the need for external interventions or special additives. Healing contributes to a longer service life of pavement, and it is commonly incorporated into the pavement design procedure by using lab-to-field factors. Such an approach is often too simplistic and based on past experiences with traditional materials. Advanced approaches are needed for a better and more reliable assessment of self-healing properties, especially for unconventional and innovative pavement materials. This requires an adequate understanding of healing phenomenon, including the relevant aspects at physic-chemical level and the driving mechanisms acting at different scales. This chapter explores the fundamental principles and recent developments in the study of intrinsic self-healing of asphalt pavements.

G. Leegwater
Netherlands Organisation for Applied Scientific Research (TNO), NEXT Delft, Delft, The Netherlands

F. Hammoum (✉)
Materials and Structures Department, Gustave Eiffel University, Campus of Nantes, France
e mail: ferhat.hammoum@univ-eiffel.fr

A. Tabaković
School of Civil Engineering, University College Dublin, Belfield, Ireland

Revolution Road Ltd, Bray, Wicklow, Ireland

O. Baglieri
Department of Environment, Land and Infrastructure Engineering, Politecnico di Torino, 24, c.so Duca degli Abruzzi, 10129 Turin, Italy

H. Baaj
Department of Civil and Environmental Engineering, University of Waterloo, 200 University Ave W, N2L 3G1, Waterloo, ON, Canada

© The Author(s), under exclusive license to Springer Nature Switzerland AG 2025

H. Baaj and O. Baglieri (eds.), *Crack-Healing of Asphalt Pavement Materials*, RILEM State-of-the-Art Reports 45, https://doi.org/10.1007/978-3-031-99443-2_2

1 The Relevance of Self-Healing for Asphalt Pavement Performance

One of the major deterioration modes affecting the service life of asphalt pavement is fatigue cracking due to repetitive loading of traffic. It is well recognized that part of the fatigue damage caused by loading is restored during rest periods between loading repetitions. Therefore, not only is fatigue damage relevant when assessing the service life of asphalt mixtures, but also the ability of the material to reduce this damage during rest periods. This ability is commonly referred to self-healing.

For this reason, research on self-healing is a relevant topic in asphalt pavement engineering. This chapter focuses on intrinsic self-healing, which is defined as the ability of an asphalt material to autonomously repair (micro)-cracks without the need for external interventions or additives [1]. The intrinsic healing properties of asphalt mixes are linked with their viscoelastic nature. Asphalt binder, the primary binding material in these mixes, exhibits unique rheological characteristics that allow the material to flow and adapt at the molecular level in response to damage. This inherent ability to self-repair micro-level fissures and imperfections forms the basis of the healing process.

However, the mechanisms behind the healing of asphalt have not been fully understood. As a result, a standardized healing test that can predict the healing behaviour of an asphalt mixture is not available yet [2, 3]. Therefore, it is not possible to irrefutably demonstrate that new materials are better or worse healers compared to existing ones. This results in two challenges. The first challenge is that the healing ability of newly developed materials, which are, for instance, more environmentally friendly, are unknown. As a result, existing design rules for healing cannot be applied to these new materials, which presents a hurdle for large-scale application. The second challenge is that it is not possible to demonstrate the effectiveness of newly developed materials that are engineered to demonstrate improved self-healing behaviour.

In asphalt mixes, different cracks can manifest at various levels, ranging from macroscopic scale (asphalt mix) to mesoscopic scale (mortar)/microscopic scale (mastic/binder), each presenting its distinct characteristics and implications. By delving into the principles and mechanisms underlying the healing process at a different scale, the aim is to shed light on the transformative potential of healing in restoring the original properties of asphalt mixes. Ultimately, understanding the healing capabilities of these materials holds the premise of longer-lasting, more resilient, and environmentally sustainable asphalt pavements.

2 The Incorporation of Self-Healing in the Design of Pavement Structures

The most important criterion in the design of pavement structures is the resistance to fatigue cracking. The repetitive loading of heavy vehicles causes tension at the bottom of the pavement, which can lead to failure. Therefore, the resistance of a material to fatigue loading is input for the design calculation. The main question asked here is "How do we consider the phenomenon of the recovery (linked to rest periods) in pavements design method?".

During laboratory fatigue tests, the tested specimen is subjected to continuous cyclic stress or strain signals. However, in actual traffic conditions, the time between the axles of heavy vehicles is typically followed by a so-called rest period. Some researchers proposed to modify the protocol of the standardized fatigue test by inserting a rest phase between each loading phase. The modified protocol highlights a gain in the life duration of asphalt specimens tested under alternating loads.

Fatigue-based mechanistic pavement design methods are widely used to determine the optimal thickness of asphalt pavements. This approach relies on calculating the tensile strain at the bottom of the asphalt layer due to the stresses imposed by traffic loads. The resulting tensile strain value serves as a key input in fatigue analysis, determining the pavement's resistance to repetitive loading cycles before failure occurs. To establish a robust fatigue relation for each mixture type, laboratory fatigue tests are conducted, generating experimental data as input for the design process. If the computed number of load repetitions to failure falls short of the required design threshold, an adjustment in the design is needed. Typically, this involves increasing the pavement thickness to reduce the strain level ε and consequently extend the pavement's service life.

It has been shown that self-healing significantly increases the fatigue life of pavement [4–10], and is one of the important explanations for why the fatigue performance observed in the lab is less compared to practice [11].

To take this positive effect into account and prevent the over-dimensioning of pavement structure, healing has been incorporated into most pavement design methods by the introduction of a healing factor [10].

Notably, in the Shell Pavement Design Manual [12], a "shift factor" is introduced to consider the influence of healing properties and the lateral dispersion of traffic (where not all wheels follow the same track). In the French pavement design method [13], a shift factor is introduced to consider the self-healing contribution in the calibration coefficient. This method consists of adjusting the results obtained in the laboratory with fatigue tests and the development of cracks observed in-situ. The asphalt's fatigue life used in practice can be listed as follows:

$$N_{field} = N_{lab} * SF \tag{1}$$

where,

SF: shift factor due to some effects (testing method, healing, residual stresses, resilient dilation, etc.).

A summary of healing and general shift factors proposed in literature is provided by Van den Bergh [10] and reproduced in Table 1. The table shows that the healing factors in design guidelines range from 2 to 20.

A review of the different pavement design methods over the world reveals two different methods to take into account the self-healing effects:

(a) Shift factor obtained from the experimental testing in the lab: [17] expressed that the shift factor is dominated by healing and is a function of the number of rest periods, the time between load cycles, and the rest period duration. The strong variation in healing factors per country or method can be explained by a combination of reasons. Firstly, the healing factor is merely an experimentally determined fudge factor that describes the difference in test results with and without healing periods. Secondly, the determined healing differs per test method used [5]. Thirdly, every design procedure prescribes a different method to assess fatigue and healing.
(b) Shift factor obtained from the full-scale validation: In the French pavement design [13], the correction factors k_c which is associated with the healing contribution varies between 1 and 1.3, and is a fitting factor based on full-scale validation. This correction factor k_c is converted to equivalent shift factor SF by the relation SF $= (k_c)^5$ and varies between 1.6 and 3.7. The full-scale validation introduces real conditions with rest periods between axle loads.

The healing behavior of airport pavements, as discussed in the literature, differs from that of conventional road pavements, particularly in terms of the required rest period for recovery: (i) the spacing of twins and aircraft landing gear differs from heavy trucks. (ii) Landing and take-off speeds are no longer of the order of those

Table 1 Healing and general shift factors according to different design procedures [10]

References	Type of test	Load mode[a]	Individual factors			General shift
			Healing	Crack propagation	Traffic wander	
Asphalt Institute						13
[12]	Variable	D	2–10	1	2.5	10–20
[13]	2-point bending	D				1.6–3.7
[14]						10–13
[15]	2-point bending	F	7.1	3	1–2.5	7.1
[16]	4-point bending	D	4			

[a]F = Force controlled; D = Displacement controlled

encountered on the road. (iii) The time intervals between aircraft movements are longer than those between two heavy vehicles. Based on some calculation, Marchand [18] concluded that the equivalent shift factor for airfield pavement varies between 4 and 5.

3 Basic Observations on the Healing Phenomenon

In the first period of healing research from 1960 to 1980, healing was often studied as a part of fatigue research [19–23]. These researchers showed that a short rest period after each loading cycle resulted in significantly longer fatigue lives. During this period, it has been hypothesized by some of the researchers that the mechanism behind healing is not actual material healing but simply viscoelastic recovery.

The first overview of healing research was presented in 1982 [5]. The work up until this point is summarized, and trends are reported and verified on asphalt mixtures. The observed trends are the following:

- All studies showed a significant amount of healing when rest periods were introduced. This is well illustrated in a figure [24], reproduced here as Fig. 1. On the horizontal axis, Tr/Tw is the ratio between loading and rest periods. On the vertical axis, Nr/Nw is the ratio of the fatigue life with rest periods to fatigue life without rest periods. As a result of intermitted healing tests, healing factors were observed from 1.5 up to 20.

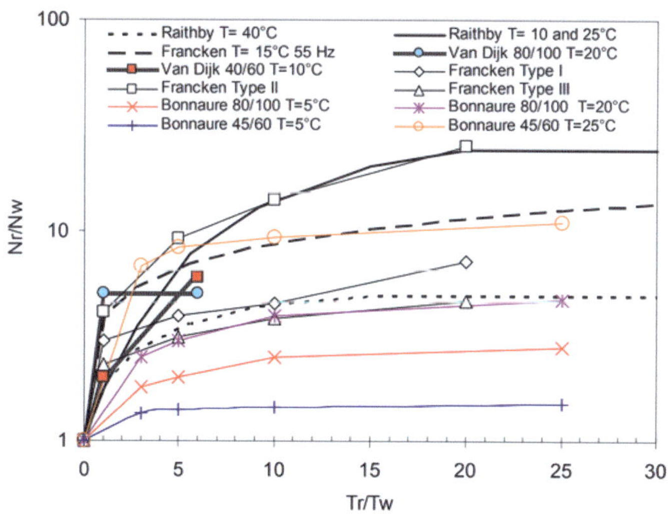

Fig. 1 Overview of reported healing from 1960's to 1980's, with the ratio of number of loads to failure with and without rest periods on the vertical axis and on the horizontal axis the rest to load ratio [24]

- Healing increases with increasing resting time. Initially, the increase in healing, with increasing healing time, is steep. However, there seems to be an upper limit to the effect of rest periods on the number of loads to failure. This can also be observed from the results presented in Fig. 1. The figure shows that even with very long rest periods (a rest-to-load ratio of 20 or more), specimens under fatigue loading will eventually fail.
- For all but one research, the amount of healing increased as the temperature increased (tests are performed between − 5 to 40 °C).
- Asphalt mixtures with a lower stiffness, usually manufactured using a softer binder, show a higher healing capacity.
- Healing tests performed under stress-controlled conditions result in more healing compared to healing tests under strain-controlled conditions when similar damage levels are considered.
- No general quantitative relations can be formulated to relate the amount of healing to the loading conditions and materials used.

4 Physicochemical Aspects of Self-Healing

4.1 Chemical Drivers for Intrinsic Self-Healing

Different research groups pose different theories behind the healing mechanisms of asphalt. Kim et al. [25] start by observing that more work is needed to open a crack after a rest period. They indicated that this is partly due to viscoelastic recovery and partly due to chemical healing. In the article, chemical healing is investigated further. The article refers to mechanisms behind the healing of polymers. It takes this as a starting point for reasoning how asphalt materials might heal [26]. To validate this hypothesis, the healing indexes of three different asphalt binders were analyzed and correlated to two parameters that indicate chain length (methylene (CH_2) to methyl (CH_3) ratio) and extent of branching in asphalt binder (MMHC ratio). The results are presented in Fig. 2, from which it can be argued that that longer chain lengths and less branching increase the healing potential.

Another perspective on the driving mechanism for healing relates to surface energy. The concept of surface energy is used for problems of cohesion and interfacial adhesion. In 1989, Schapery introduced surface energy density to explain cracking and self-healing processes. Later, Lytton et al. [27] offered an interpretation of asphalt binder cracking and self-healing based on Schapery's theory and used the concept of surface energy density. The surface energy density essentially depends on the contents of different components of asphalt binder (aromatic, amphoteric and paraffinic).

The surface energy of asphalt binder is obtained from wetting and dewetting angle measurements using the Wilhemly plate. The wetting surface energy is directly related to the self-healing potential.

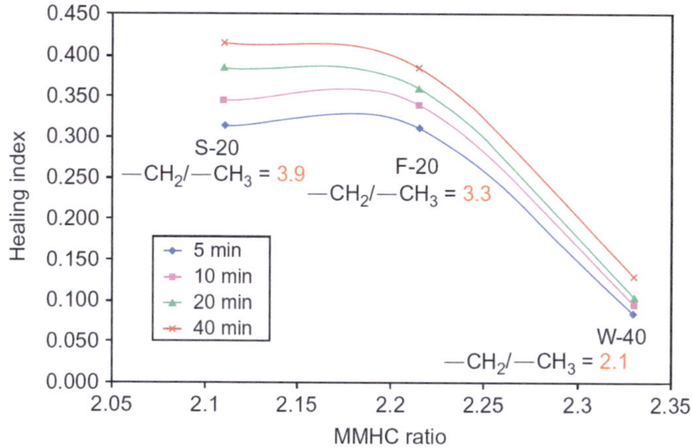

Fig. 2 Correlation of MMHC ratio and healing index [7]

Kim et al. [28] and Little et al. [29] conducted mechanical tests to evaluate the self-healing of five asphalt binders at the specimen scale (combined effect between wetting and inter-diffusion). After a cyclic loading phase, a rest phase is introduced, and several rest times are tested. A pseudo-strain energy obtained from the area under the stress–strain curve is used to estimate the damage in the specimen at each cycle. The self-healing parameter (healing index) H is calculated after each rest phase as follows:

$$H = \frac{\varnothing_A^R - \varnothing_B^R}{\varnothing_A^R} \tag{2}$$

with $\varnothing_A^R$ and $\varnothing_B^R$ respectively representing the strain pseudo-energy after and before a rest period measured during the cyclic loading test. The results are presented in Fig. 3.

According to the work of [29], the short-term self-healing kinetics in asphalt binder (the first 10 s of the rest period) is inversely proportional to the non-polar component of surface tension or Lifshitz-van der Waals parameter.

Similarly, the self-healing kinetics (after the first 10 s) is directly proportional to the polar component of the surface tension of the asphalt binder or Acid–Base component. These two components of surface energy for wetting and dewetting are identified by [30]: Lifshitz-Van der Waals and Lewis Acid–Base. In the literature, these two components explain cohesion and adhesion in asphalt mixtures. In summary, an asphalt binder that has a high self-healing potential has maximum polar energy and minimum non-polar energy [11].

Building on this work Bhasin et al. [31], have further in investigated healing and surface energy using the two-piece healing method [31]. In this test, two pieces of asphalt binder are brought together in a Dynamic Shear Rheometer (DSR), and the measured stiffness in shear of the combined two pieces is compared to the response

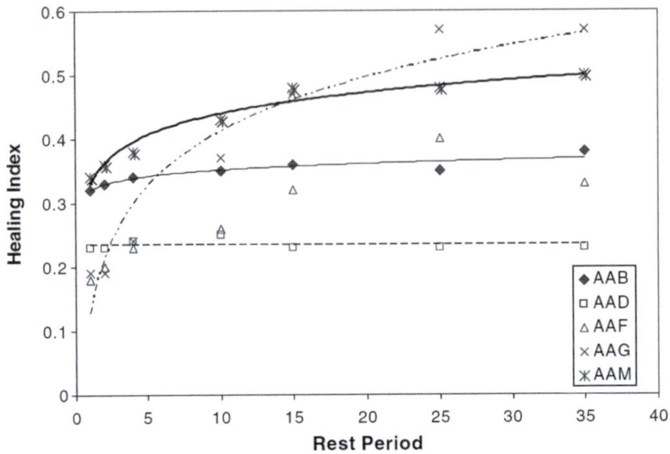

Fig. 3 Intrinsic self-healing of asphalt mixes evaluated with different binders [29]

of a single piece of asphalt binder. At the start of the test, two pieces of asphalt binder are pushed together. It is assumed that the total area is instantly in contact. Measured healing is assumed to consist of an instant part that can be related to surface energies and a time-dependent part that is linked to diffusion. The ARC quarterly report of April to June 2008 [32] presented a graph showing the measured healing of five different binders from the SHRP program, as shown in Fig. 4.

They propose the following formula (Eq. 3) to describe intrinsic healing:

$$R_h(t) = R_0 + p\left(1 - e^{-qt^r}\right) \tag{3}$$

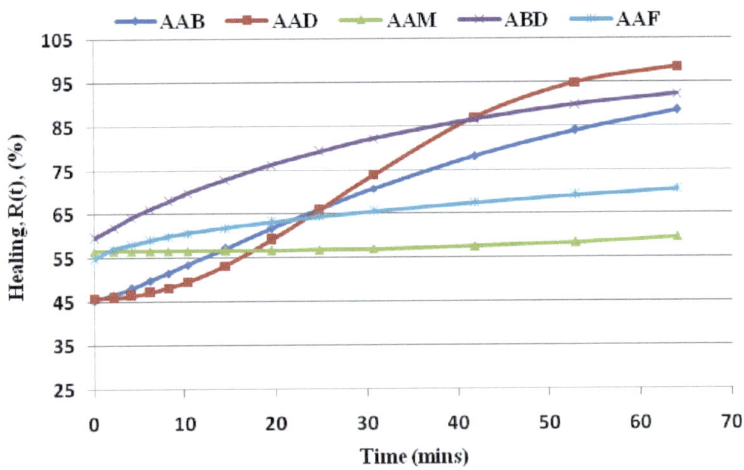

Fig. 4 Healing measured with two-piece healing methods [32]

where R_0 is the instant healing based on the material's surface energy, the other part is the time-dependent part. The R_0 levels correlate very well with the surface energies determined by the Wilhelmy Plate Method [7].

Qiu [33] also performed tests using this two-piece healing method. He has demonstrated that the pressure level during the tests influences the shear resistance measured. This influence of pressure could imply that part of the shear resistance that is measured is caused by friction. When the two pieces of asphalt binder are pressed together, the surface roughness of the asphalt binder samples will influence the amount of shear resistance that is measured. Even though asphalt binder looks smooth, Atomic Force Microscopy (AFM) images show that on an extreme scale, asphalt binder indeed has surface texture [34].

The link between chemical composition and the healing potential of binders has also been highlighted by several other authors [35, 36]. They related stiffness recovery of binders in DSR fatigue tests to, amongst others, SARA analysis. These tests demonstrate that the healing potential of binders increases when more small molecules are present in the asphalt binder. They show that as the Saturates/Aromatics ratio increases, the observed healing also increases.

More recently, it has been demonstrated by AFM that micro-cracks form and disappear in the stress concentration zone [37, 38]. Also, by using AFM, it has been found that additives can affect the rate of crack closure in asphalt binders [39]. Furthermore, Song et al. [40] showed self-healing in asphalt mastics through a reduction in the percentage of voids after rest periods using X-ray computer tomography.

4.2 *Influence of Thermal Expansion Self-Healing*

For the case where physical cracks are present, the meso to macro scale, Grossegger [41] shows that thermal expansion of asphalt materials, e.g. asphalt binder and aggregates, is one of the main factors influencing asphalt self-healing. According to Grossegger, the self-healing process follows these steps, which have been summarized in Fig. 4:

1. When a crack opens in asphalt, the surface of the cracks is smoothened due to asphalt binder's surface tension.
2. Asphalt may heat up due to solar irradiation; if that occurs, asphalt binder and aggregates may expand, which brings the cracks together and creates contact points.
3. Due to surface tension, gravity, and further thermal expansion, mastic may flow into the cracks and fill them from these contact points.
4. The material that fills the crack is a mixture of asphalt binder and fine aggregates; the fine content and composition may change the expansion coefficient and viscosity, affecting how cracks are filled.

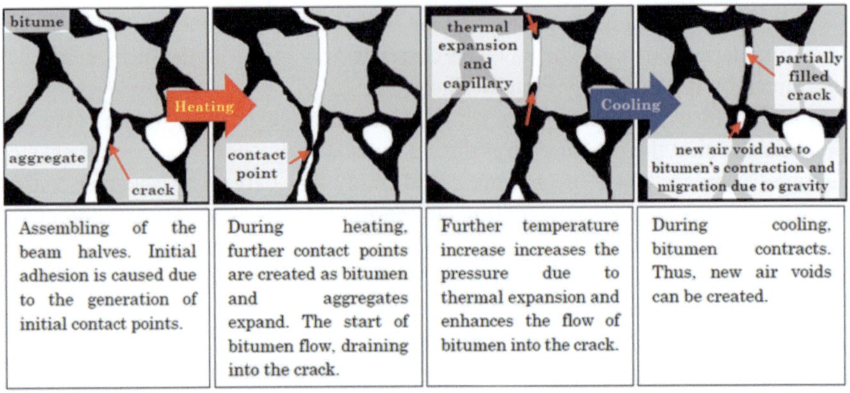

Fig. 5 Scheme of the natural self-healing process [43]

5. During heating, cracks will continue filling until a thermal equilibrium is reached. When the thermal equilibrium is reached, asphalt binder will drain due to gravity, which may open additional voids in the areas close to the surface. A similar effect was noted by [42] that used induction heating to accelerate the mending of cracks, once the asphalt reached the thermal equilibrium, new pores opened in the asphalt, and its strength was reduced; regardless of the high temperature of the asphalt.

Furthermore, Grossegger et al. [43] reported that the presence of water in the cracks hindered the self-healing process due to water having a similar density to that of asphalt binder, preventing the drain of asphalt binder into the cracks, as presented in Fig. 5.

Based on this, it can be deduced that aggregates and binders with higher thermal expansion coefficients will accelerate asphalt self-healing. Additionally, using asphalt binder with lower viscosity can accelerate the healing process, as it allows the binder to flow more easily into cracks, facilitating repair. The amount of asphalt binder available to flow and drain into the cracks will influence the asphalts self-healing, which may be why porous asphalts shows higher self-healing capacity than denser materials [44]. The self-healing of cracks in asphalt derives from the complex interplay between asphalt binder viscosity, the thermal properties of aggregates and asphalt binder, and the aggregate gradation. Based on these insights, one of the most promising measurements of the natural self-healing capacities of asphalt may derive from simply measuring its thermal expansion coefficient.

5 Healing Mechanisms at Different Scales

An asphalt mixture is a heterogeneous composite material, composed of gravel, sand, filler and asphalt binder. For the design of asphalt pavements, knowledge is required of the self-healing behaviour of full asphalt mixtures. However, studying asphalt mixtures is quite complex, as the material is heterogenic with numerous interfaces between different constituents, which also exhibit a broad range of sizes.

Mineral aggregates (gravel, sand, and filler) will not restore any strength after damage under the conditions present in pavements. Therefore, the restoration in asphalt mixtures can be assumed to originate from the asphalt binder. As a result, it is expected that the restoration behaviour of asphalt binder has predictive power for the restoration behaviour of a mixture.

It should be kept in mind that mixture composition and chemical interactions between the binder and the aggregates can influence the restoration of an asphalt mixture. For instance, Little et al. [45] showed that certain binders show more healing when combined with specific aggregate types. When damage occurs on the interface between mineral aggregates and the binder, the aggregates might also affect the interaction between the newly created surfaces. Hence, to fully assess healing behaviour, both the behaviour of the binder and the mixture is relevant. A possible approach can also be to study healing in an asphalt mortar. In this case, all the chemical interactions are present in a highly reproducible environment due to a lack of gravel reducing the anisotropy in the material.

As a result of the driving influence of the binder on healing, combined with the expected interactions with aggregate surfaces, healing research has been conducted on all scales; asphalt binder, a mix of asphalt binder, a mix of asphalt binder and filler (mortar), filler and sand (mastic) and the whole asphalt mix. Through this research executed at different scales it has also demonstrated that correlations between the different scales are not easily established [24].

Several researchers have adopted a healing model from polymer science [3, 6, 7, 31]. The polymer healing model interprets the self-healing of a crack as a series of five processes that occur subsequently: (i) surface rearrangement, (ii) surface approach, (iii) wetting, (iv) diffusion and (v) randomization [46]. These processes are visualized in Fig. 6. As a result of these processes, two mechanisms that determine the interaction between two crack interfaces have been identified. First, surfaces must be in contact with each other to be able to transfer load, this is named wetting. Second, the force that the contact area can transmit is dependent on time. As a result of adhesion, there will be an instant ability of the contact area to transfer a load. This initial strength will increase over time as molecules diffuse over the crack interface and increase interaction between the two crack phases. Therefore, it is concluded that the macroscopically observed healing is the convolution of the amount of area in contact and the load-bearing capacity of this contact.

To translate the conceptual model of healing into equations that could be used to assess healing at the macroscale, Wool and O'Connor introduced a parameter. This parameter is defined as the ratio of the material's mechanical properties after

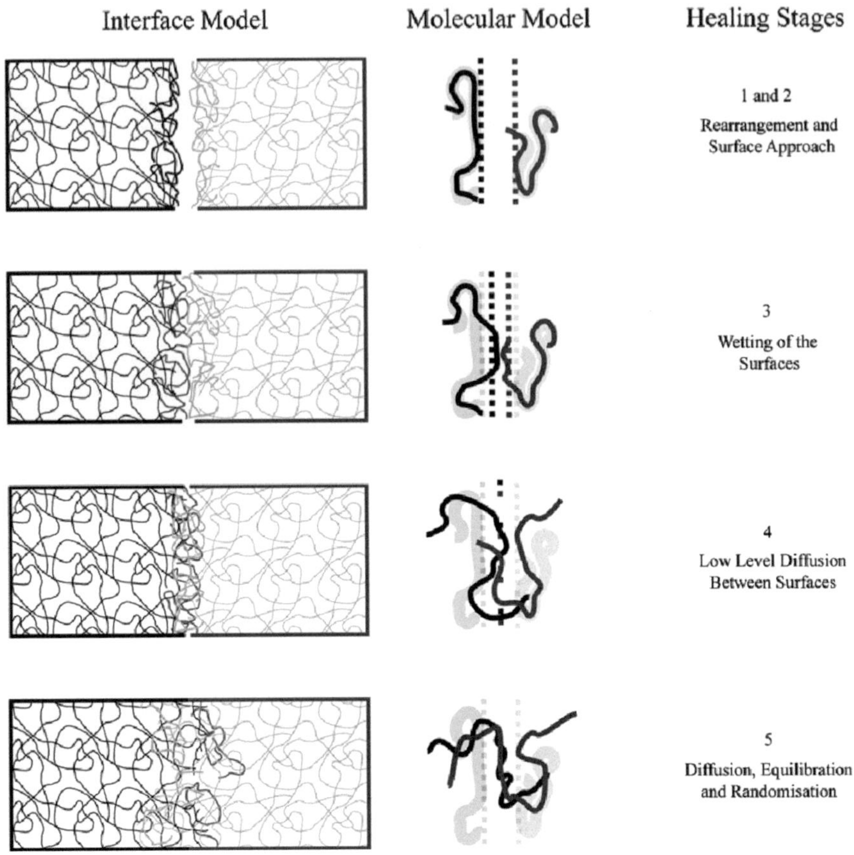

Fig. 6 Five steps in the healing process [46, 47]

a rest period to its mechanical properties in the original state. The value of R, by definition, varies between 0 and 1. These mechanical properties include fracture stress, failure strain, tensile modulus, and fracture energy. The recovery parameter, R, is represented as a convolution integral of the intrinsic healing function $R_h(t)$ and wetting distribution function $\varphi(t, X)$ given by

$$R = \int_{\tau=-\alpha}^{\tau=t} R_h(t-\tau)\frac{d\varphi(\tau, X)}{d\tau}d\tau \tag{4}$$

Here, τ is the time variable, and t is the present time. This intrinsic healing function defines the rate at which the self-healing occurs among two wetted surfaces and the portion of the original strength recovered. Hence, this intrinsic healing function is additively decomposed to include the intrinsic self-healing due to short-term wetting and the intrinsic healing due to long-term diffusion and is represented using

$$R_h(t) = R_0 + Kt^{0.25}\xi(t) \quad (6)$$

where, R_0 is the instantaneous strength gain, K is a constant, and t is time. $\xi(t)$ represents the effect of surface rearrangement of molecules. This crack healing model through inter-diffusion is further studied by Kim et al. [48], who conducted studies based on crack healing theory at the polymer–polymer interface. While this model has been validated for polymers, the extent of validation for asphalt binder is limited at this moment [3, 7]. It is also likely that the model should be adjusted to serve as a model for asphalt binder, as the types of molecules present in asphalt binder differ from polymers.

6 Closing of Cracks

Stiffness and strength loss in asphalt materials is a consequence of the formation of micro cracks. To heal this real damage, the two sides of the micro crack must regain contact. This process requires the displacement of material into the crack to ensure proper wetting. Following this, the contact area can homogenize to reestablish molecular bonds, restoring the material's strength and toughness (see Fig. 7). Asphalt binder, being a very viscous liquid, can flow inside a crack and reinstall contact above its glass transition temperature (T_g). There can be variety of forces triggering this flow, such as an external applied compressive force, gravity, a compressive force as a result of temperature expansion, surface energy, etc. As the flow properties of asphalt binder are very temperature-dependent, this will happen faster at higher temperatures. In intrinsic healing, this flow is the natural flow of the binder, while in the case of extrinsic healing, attributes are brought into the material or the system to speed up this process.

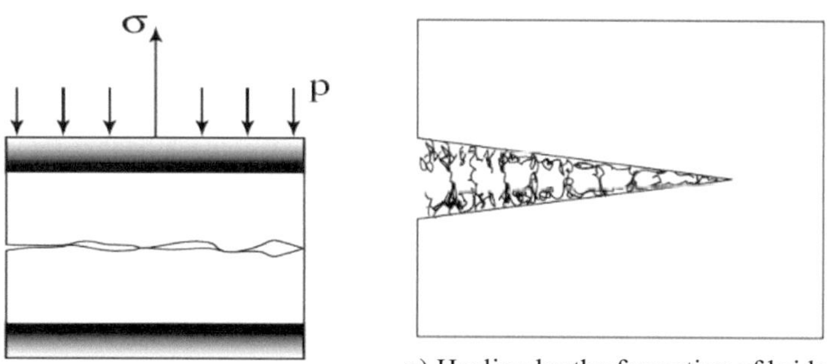

a) Bringing the lips of the crack in contact

a) Healing by the formation of bridges between molecular chains

Fig. 7 Principle of intrinsic self-healing of a crack

The crack can close fully or partially, depending on the time given and the effectiveness of the self-healing process. The extent of the closing of the crack has a strong influence on the observed material response. Figure 8 presents an example of a crack, which appears to have disappeared after a period without loading.

Figure 9 shows microscopic images of the self-healing of a fully cracked dog-bone sample made of asphalt mastic while it is kept under controlled conditions. It demonstrates that after a healing period of 18 h, the original crack is hardly detectable using this technique. In his work [24], also proves that the strength of these samples restores over time.

In relation to the crack closure at the micro and macro levels, it has been demonstrated that the initial properties of asphalt concrete cannot be fully recovered only using heating [50]. The authors concluded that healing through heating cannot

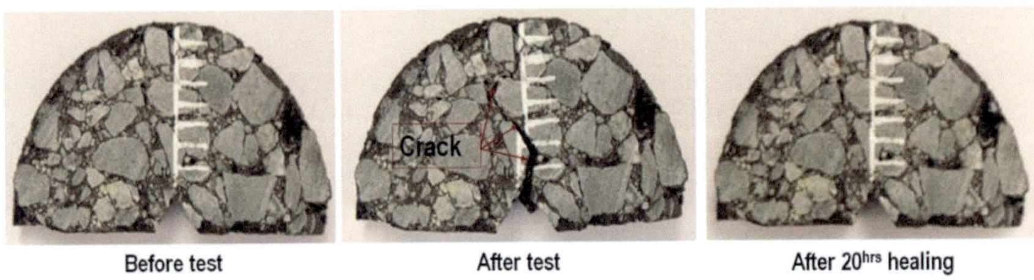

Fig. 8 SCB test specimen crack closure/healing [49]

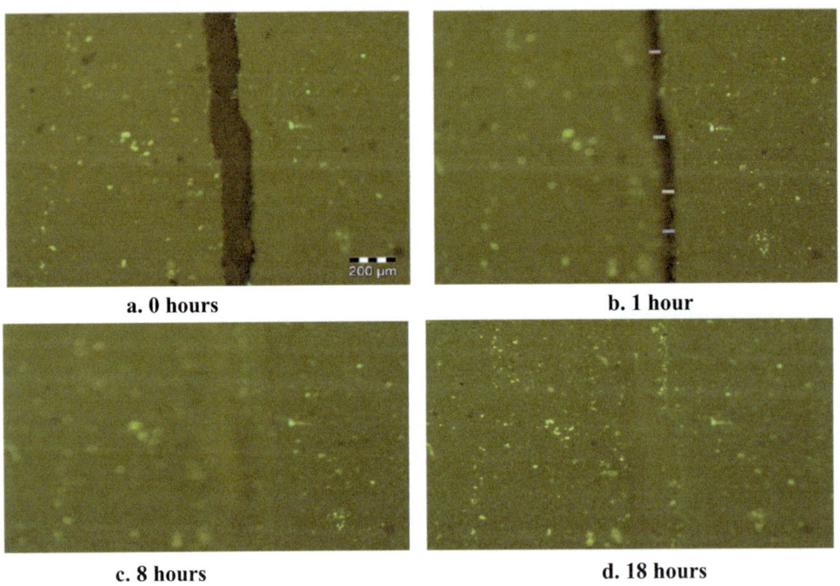

Fig. 9 Fluorescence microscopic pictures of an asphalt mastic specimen at a healing temperature of 25 °C with different healing times [24]

recover all types of cracks, as in case of significant plastic deformations, the binder flow is not extensive enough to re-establish the contact between aggregates. Consequently, although the cracks were closed (crack damage was repaired) due to the binder flow, the stresses generated in the area close to them were higher than those in the rest of the material.

There are several tests developed to assess the healing ability of cracks in asphalt binders. [51, 52] propose applying a load on an asphalt binder sample in a plane/plane shear configuration using a conventional Dynamic Shear Rheometer (DSR) test. The test would be conducted at a specified frequency and under low levels of deformation until the stiffness of the specimen reduces to half of its initial, virtually undamaged value obtained at the start of the test. Then, they introduce a rest phase (T_R) of variable duration before loading the sample again (Fig. 10).

Figure 11 shows that the insertion of rest phases, as well as the increase in their duration compared to the duration of the stress, makes it possible to increase the life duration of the specimen. [52] deduce a favorable effect of the duration of the rest phase on the self-healing of the asphalt binder.

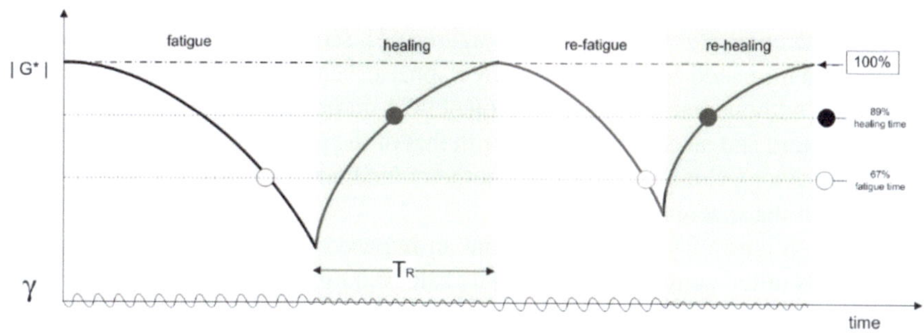

Fig. 10 Insertion of rest period between loading phases [51]

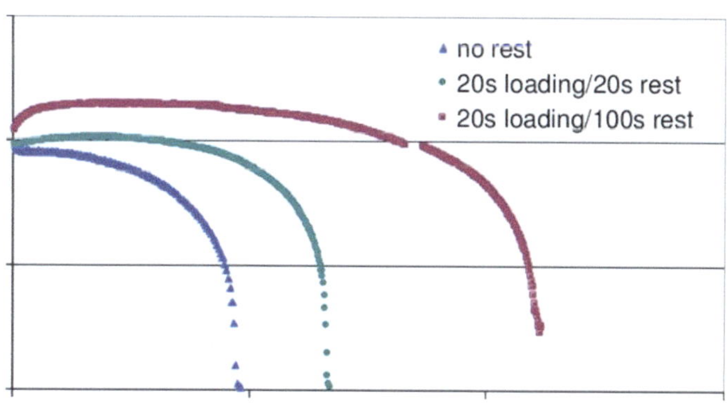

Fig. 11 Effect of rest periods on fatigue tests (DSR) of Asphalt Binder (unaged) at 210 kPa stress, 25 Hz and 15 °C [52]

In their work, Bhasin et al. [53] state that the growth of a micro-crack is linked to the creation of new fracture surfaces originated by accumulated damage in the vicinity of the crack tip. This damage is related to the deformation and rearrangement of molecules in this failure zone. The authors hypothesize that the self-healing mechanism could be seen as the reversal of crack initiation and propagation. To explain this mechanism at a micro-scale, a theoretical approach was proposed where fracture mechanics were used to model the micromechanics of crack growth and self-healing in asphalt composites by quantifying time-dependent healing. Two distinct phases across a crack interface are differentiated: wetting and intrinsic healing (or rate of strength gain processes). According to this model, wetting is related to the material's mechanical properties and surface-free energy. In contrast, intrinsic healing is considered a time-dependent phenomenon associated with the self-diffusion of asphalt molecules across the crack interface. It was experimentally proved that the rate of crack closing or wetting rate depends on the crack geometry, mechanical properties of the binder or mixture, and the surface free energy of the binder.

At the University Gustave Eiffel, the repeated local fracture test on asphalt binder is the device used to study cracking and self-healing behavior [54]. This device consists of two steel sample holders whose central protuberances, when placed face to face, represent the contact zone between two aggregates, see Fig. 12. Each protuberance consists of a truncated cone terminated by a spherical cap. This geometry ensures, in a specimen without crack, a concentration of 90% of the stresses in a cylinder with a radius of 3 mm and an axis coincident with that of the protuberances (spherical part of the test device). Consequently, this geometry facilitates the initiation of a crack in the center of the specimen.

After each tensile cycle obtained from an imposed displacement, the specimen returns to its initial configuration. To study self-healing, the rest time is fixed after a loading phase, and an ultrasonic technique is used to monitor the reattachment phase of the lips of the crack.

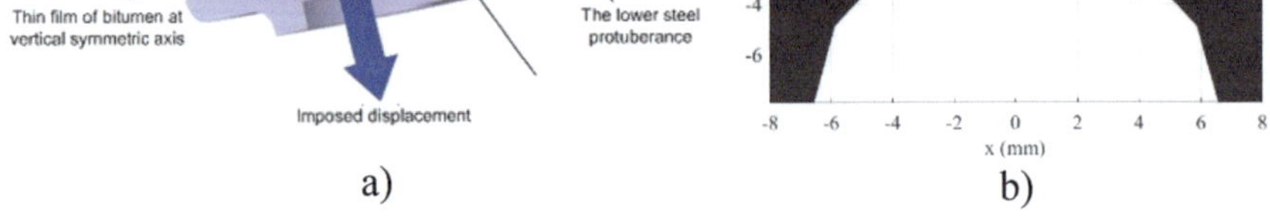

a) b)

Fig. 12 The local fracture test: **a** test device and **b** sample geometry [55]

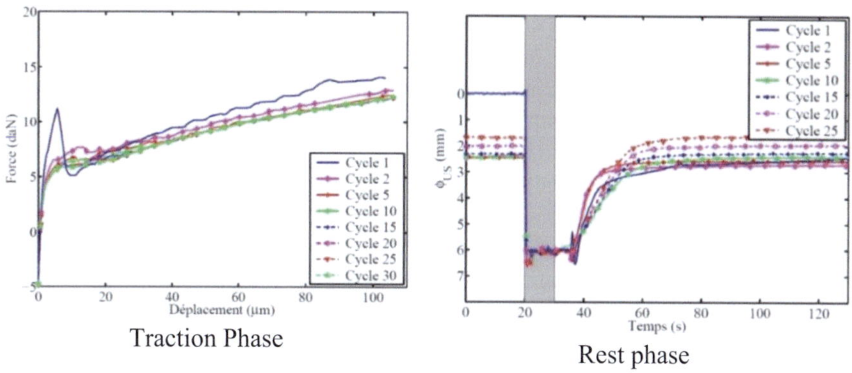

Fig. 13 Self-healing of modified asphalt binder by polymer at 0 °C—local fracture test [56]

This device made it possible to study the kinetics of reattachment at the edge of a crack during the rest phase. Experimental tests show that kinetics depend on experimental conditions and the nature of the used asphalt binder [56], see Fig. 13.

Care must be taken when interpreting results from tests conducted at low strain levels and/or long rest periods, as changes in material properties may be attributed to either nonlinear behavior or physical hardening of the material.

7 Biasing Effects in Fatigue and Healing Testing

Healing tests can be executed using many approaches. When self-healing is measured after a period of fatigue loading, the stiffness recovery after a rest period is often close to 100%. This can be observed in the example shown in Figure. Next to this case, there are many other reports of this phenomenon [57–60]. However, despite this strong restoration of stiffness after a healing period, the number of loading cycles to failure is often significantly reduced compared to the original performance. A plausible explanation for this phenomenon is that asphalt materials lose stiffness in repetitive loading before any damage in the material occurs. This specific loss of stiffness is due to phenomena other than damage, and therefore is fully reversible.

In order to avoid gross errors in estimating the fatigue and healing performance of materials, it is thus very important to isolate transient and reversible non-fatigue phenomena that occur during repeated loading [61–63]. The concept of "biasing effects" have been proposed to define this type of phenomenon. The biasing effects generally treated in the literature are non-linearity, heating and thixotropy.

Non-linearity refers to the phenomenon where the response of asphalt material depends on the level of strain applied. If the strain is different during the test, the detected stiffness also varies. Heating of the test specimen occurs because of energy dissipation during cycling loading. As the response of asphalt materials is very temperature-dependent, a small increase in temperature will result in a loss of stiffness.

The phenomena behind thixotropy requires a bit more explanation. According to the "International Union of Pure and Applied Chemistry" (IUPAC), the application of finite shear stress to a system after a long period of rest can cause a decrease in its viscosity or consistency. If this decrease persists when the stress is interrupted, the behaviour is called "work softening". At the same time, if the original viscosity or consistency is recovered, the behaviour is called "thixotropy" (IUPAC website). The accepted explanation of thixotropy is linked to the hypothesis of a microstructure that slowly forms in the material at its molecular level due to the weak forces of attraction between the particles. The concept of intermolecular forces as a driver for asphalt binder behaviour has already been suggested by chemists who studied asphalt binders [64, 65]. Considering the nature of these intermolecular and the large molecules composing the asphalt binder, it seems very plausible that ordering over time is likely. In this sense, it is also very possible that the application of a flow (shear) causes rupture and reorganization of this network between the particles [66]. Experimental evidence of thixotropic behaviour has been found for asphalt materials [67–70]. It has been shown that thixotropy can have a major influence on the variation of the mechanical properties of asphalt mastics during fatigue tests compared to heating [71]. The same observation was made for asphalt mixes [72, 73].

To study these biasing effects, ENTPE has developed a tension–compression test protocol called ALFABET, "Advanced Laboratory Fatigue and Biasing Effects Test") [74–77]. This test approach makes it possible to isolate different phenomena that contribute to stiffness loss and recovery during laboratory fatigue tests. Figure 14 shows the results of stiffness development during five consecutive load and rest cycles. Each load cycle entails 100,000 load cycles, and low-strain measurements are executed during the rest period to assess the recovery. Using this approach, the stiffness loss is assigned to various biasing phenomena and a recoverable component. The figure shows that the unrecoverable part is relatively minor compared to the biasing effects. This observation is confirmed for all five damage and restoration periods in Fig. 15.

The results of the last fatigue stage of the ALFABET test were analyzed to determine the fatigue life of the asphalt mixes. Four criteria were used: DER (Dissipated Energy Ratio), maximum of the phase angle and local criteria, related to the relative deviations of the deformation amplitude ($\pm$ 25%) and the phase angle ($\pm$ 5°) between the three extensometers. The fatigue life Nf was determined as the mean of the values of the four lifespans obtained with each criterion. The variations of $|E^*|$ and φ throughout the test were analyzed in terms of damage. The DIIIC approach, developed by a research team from ENTPE [78–81] was applied to correct the results of the ALFABET trial for biasing effects. The results obtained were compared with the estimates of the biasing effects made during the five fatigue loops and the corresponding rest periods, with both analyses give consistent results. The corrected values do not show a significant increase in damage during the first four fatigue stages, while they indicate a rapid deterioration of Liner Viscoelastic (LVE) properties during the

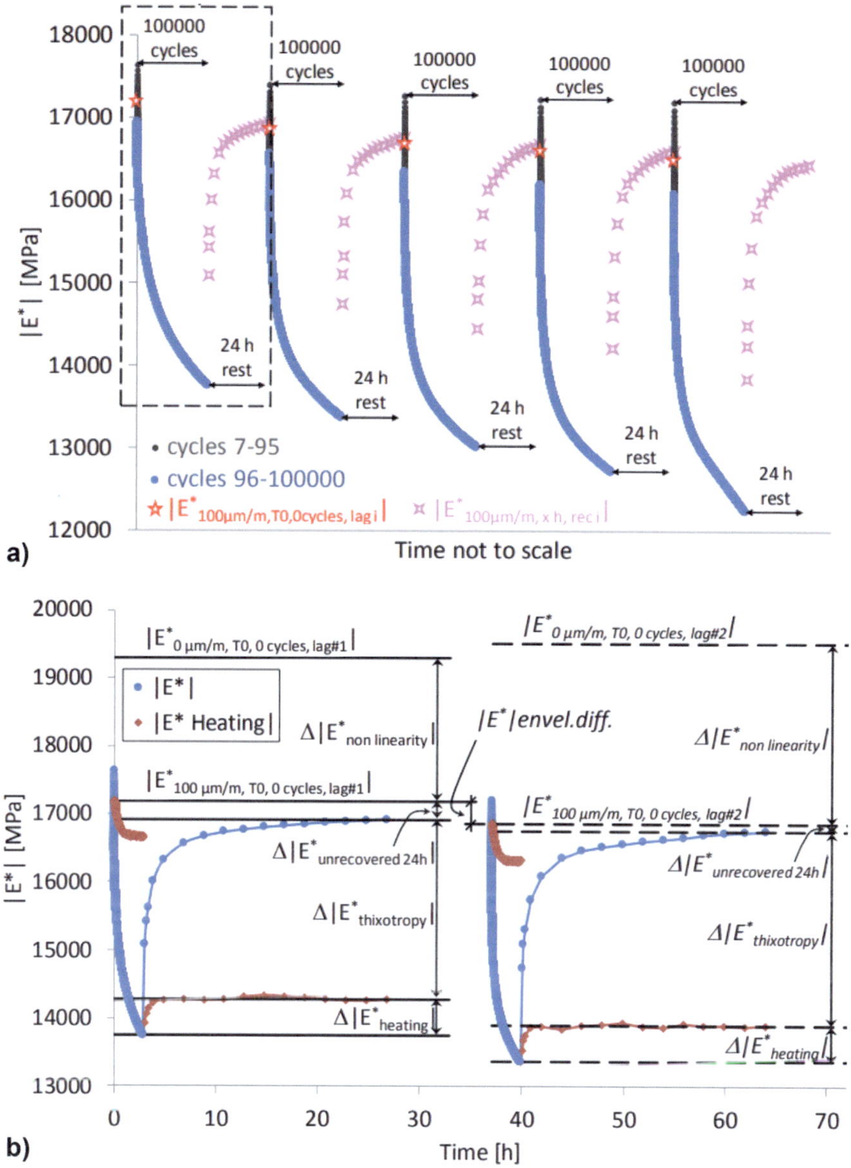

Fig. 14 Example of results obtained with the ALFABET test: (**a**, top) evolution of |E*| during loading and rest periods of fatigue test; (**b**, bottom) quantification of biasing effects and unrecovered variations during cyclic loading and rest periods #1 and #2

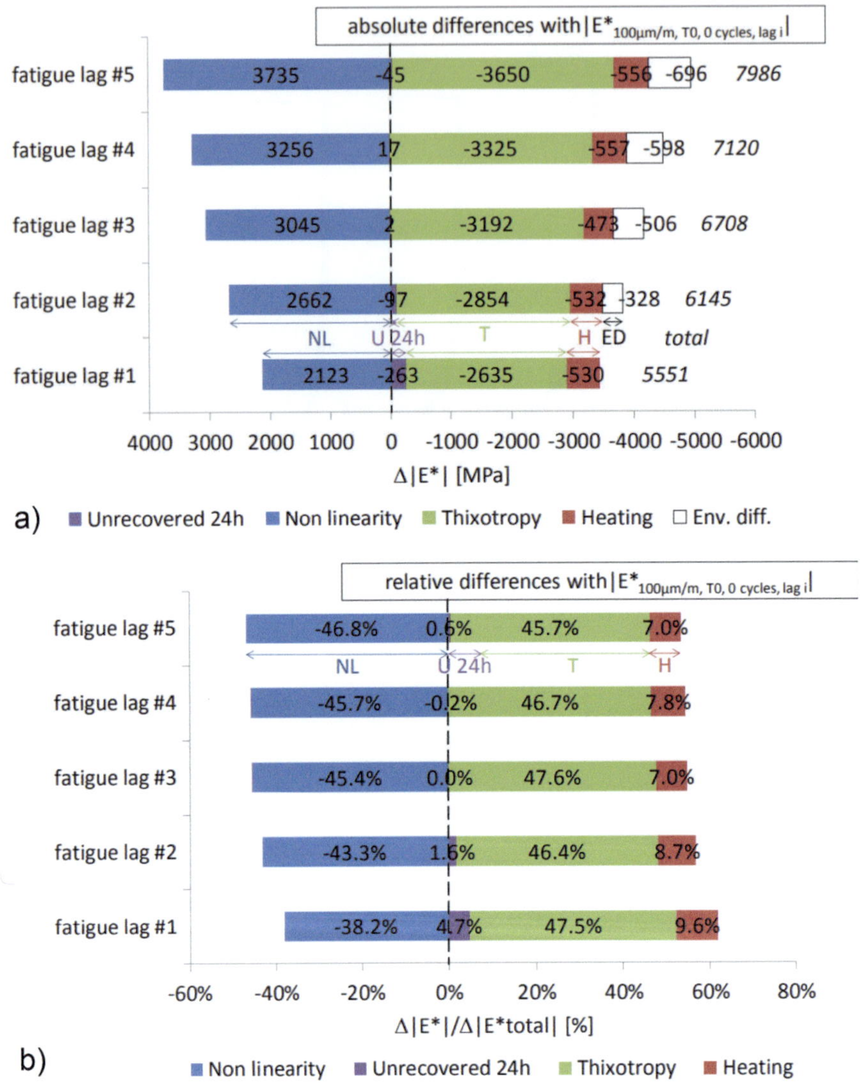

Fig. 15 Example of quantification of bias effects and unrecovered variations of |E*| obtained with the ALFABET test: (**a**, top) variations of |E*| in absolute terms; (**b**, bottom) relative variations to total variation (per period of cyclic loading)

fifth and especially the final fatigue stage. This demonstrates that the loss in stiffness is not relevant for true fatigue failure.

8 Terms and Definitions on Self-Healing and Restoration

To clarify the biasing effects (such as thixotropy, local heating, and non-linearity) as well as actual damage and healing, specific terms and definitions have been developed in relation to crack-healing [1]. As these definitions will aid in having a specific discussion about the mechanisms, the definitions and a visual representation are reproduced here. The total restoration of material properties after healing involves the recovery of performance due to the reversal of biasing effects and the self-healing of (micro) cracks within the material (see Fig. 16). The related definitions are presented below the figure.

Damage: loss of original mechanical properties, i.e. the strength or the stiffness, due to the initiation, coalescence, and propagation of micro-cracks within the material.

Restoration: The total observed change in mechanical properties after a period of rest.

Recovery: Component of the restoration that can be attributed to changes in response resulting from cyclic loading, more specifically heating and thixotropy.

Self-Healing: Component of the restoration that can be attributed to the closure and repair of (micro) cracks.

Intrinsic (Self-)Healing: Aspect of the self-healing behaviour that is inherent to the material used.

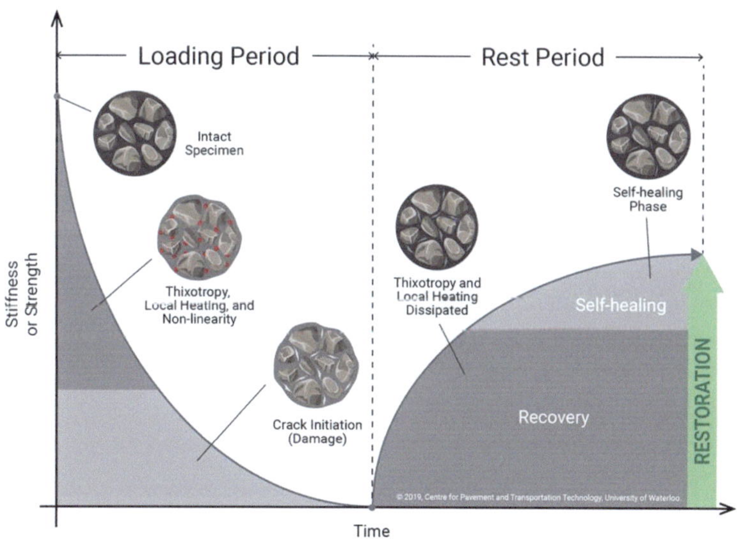

Fig. 16 Schematic overview of contributions to the restoration of performance of different phenomena [1]

Extrinsic (Self-)Healing: Aspect of the self-healing behaviour that can be attributed to an added phase or action that is specifically used to improve self-healing capabilities, to trigger crack-healing or prevent crack propagation.

9 Summary

The recovery strength of specimen after a rest period has been observed in many situations. This is a widely recognized property of asphalt materials and is consequently integrated into many pavement design approaches. However, the behavior is not yet fully understood. This is partly caused by differences in self-healing performance when the same material is tested with different healing tests. However, there are some general trends in healing; healing increases with longer rest periods, at higher temperatures, is more efficient with smaller damages, the materials have a higher penetration grade or a higher binder content.

Theories from material science have been formulated to explain the mechanisms underlying self-healing behavior. Addressing what mechanism are at play in the closing of cracks, identifying that surfaces need to restore contact through a process of wetting and that time is required for the material at areas in contact to restore its original properties through a process of homogenization. This demonstrates that the formation and healing of cracks are continuous processes within the material, influenced by loading and rest conditions. This complexity highlights why capturing this healing behavior remains so challenging. These theories help formulate better test methods; however, they are also not fully verified. The presence of biasing effects, which cause changes in material behavior occurring during tests execution that are not related to damage, hampers identification of physical damage in the material and consequently the self-healing.

This all possess a challenge when the self-healing ability of innovative materials is needed to manage risks during application. More research on healing could help better understand the phenomenon and to formulate adequate test methods to assess healing behavior of innovative materials. The presented common set of definitions on self-healing of asphalt will contribute to a clearer scientific discussion while this research is being executed.

References

1. Leegwater, G., Taboković, A., Baglieri, O., Hammoum, F., & Baaj, H. (2022). Terms and definitions on crack-healing and restoration of mechanical properties in bituminous materials. In *Proceedings of the RILEM International Symposium on Bituminous Materials. ISBM 2020. RILEM Bookseries* (Vol. 27). Springer. https://doi.org/10.1007/978-3-030-46455-4_6
2. Pooyan, A., Moreno-Navarro, F., & Rubio-Gámez, C. (2016). The healing capability of asphalt pavements: A state-of-the-art review. *Journal of Cleaner Production, 113*, 28–40.

3. Leegwater, G., Scarpas, T., & Erkens, S. (2016). Direct tensile test to assess healing in asphalt. *Transportation Research Record, 2574*(1), 124–130.
4. Bazin, P., & Saunier, J. (1967). Deformability, fatigue and healing properties of asphalt mixes. In *International Conference on the Structural Design of Asphalt Pavements.*
5. Bonnaure, F. P., Huibers, A. H. J. J., & Boonders, A. (1982). A laboratory investigation of the influence of rest periods on the fatigue characteristics of bituminous mixes (with discussion). In *Association of Asphalt Paving Technologists Proceedings* (Vol. 51).
6. Richard Kim, Y., Little, D. N., & Benson, C. (1990). Chemical and mechanical evaluation on healing mechanism of asphalt concrete (with discussion). *Journal of the Association of Asphalt Paving Technologists, 59*
7. Little, D. N., Bhasin, A., & Darabi. M. K. (2015). Damage healing in asphalt pavements: Theory, mechanisms, measurement, and modeling. In *Advances in asphalt materials* (pp. 205–242). Woodhead Publishing.
8. Qiu, J. (2012). *Self-healing of asphalt mixtures, towards a better understanding.* Dissertation report. Delft University of Technology.
9. Miglietta, F., Shane, U. B., Tsantilis, L., Baglieri, O., & Santagata, E. (2023). Self-healing master curves of bituminous binders: A non-linear viscoelastic continuum damage framework. *Road Materials and Pavement Design, 24*(sup1), 124–144.
10. Van den Bergh, W. (2011). *The effect of ageing on the fatigue and healing properties of bituminous mortars.*
11. Williams, D., Little, D. N., Lytton, R. L., Kim, Y. R., & Kim, Y. (2001). Micro-damage healing in asphalt and asphalt concrete. In *Volume II Laboratory and Field Testing to Assess and Evaluate Micro-Damage and Micro-Healing, FHWA-RD-98-142.* Texas Transportation Institute, 182p.
12. Shell. (1978). *Shell pavement design manual—Asphalt pavement and overlays for road traffic.* Shell International Petroleum Company Limited.
13. LCPC. (1997). French design manual for pavement structures. In *Technical guide, LCPC/ SETRA, translation of the French version December 1994*, 248p.
14. Von Quintus, H. L., Scherocman, J. A., Hughes, C. S., Kennedy, T. W. (1991). *Asphalt aggregate mixture analysis system.* NCHRP report 338, March 1991. ISBN 0-309-04861-3
15. Verstraeten, J. (1974). Loi de fatigue en flexion répétée des mélanges bitumineux. *Bulletin de Liaison des Laboratoires des Ponts et Chaussées*, n° 70, mars 1974
16. Groenendijk, J. (1998). *Accelerated testing and surface cracking of asphaltic concrete pavements.* Delft University of Technology, Delft, The Netherlands.
17. Lytton, R. L., Uzan, J., Fernando, E., Roque, R., Hiltunen, D., & Stoffels, S. (1993). *Development and validation of performance prediction models and specifications for asphalt binders and mixes.* Strategic Highway Research Program SHRP-A-357.
18. Marchand, J. P. (1981). *Influence du temps de repos sur le comportement à la fatigue des matériaux de chaussées.* LCPC Report in French, 10p.
19. Raithby, K. D., & Sterling, A. B. (1970). The effect of rest periods on the fatigue performance of a hot-rolled asphalt under reversed axial loading and discussion. *Association of Asphalt Paving Technologists Proceedings.*
20. Van Dijk, W., et al. (1972). The fatigue of bitumen and bituminous mixes. In *Third International Conference on the Structural Design of Asphalt Pavements, Grosvenor House, Park Lane, London, England* (Vol. 1), September 11–15.
21. Van Dijk, W., & Visser, W. (1977). Energy approach to fatigue for pavement design. In *Association of Asphalt Paving Technologists Proceedings* (Vol. 46).
22. Francken, L. (1979). Fatigue performance of a bituminous road mix under realistic test conditions. *Transportation Research Record, 712.*
23. Verstraeten, J., Veverka, V., & Francken, L. (1982). Rational and practical designs of asphalt pavements to avoid cracking and rutting. In *Fifth International Conference on the Structural Design of Asphalt Pavements* (pp. 45–58).
24. Qiu, J. (2012). Self-*healing of asphalt mixtures.* Master Thesis, Technical University Delft.
25. Kim, Y. R., Dallas, N. L., & Benson, F. C. (1990). Chemical and mechanical evaluation on healing mechanism of asphalt concrete. *Journal of the Association of Asphalt Paving Technologists, 59.*

26. Wool, R. P., & O' Connor, K. M. (1981). A theory crack healing in polymers. *Journal of Applied Physics, 52*, 5953–5963.
27. Lytton, R. L., Chen, C. W., & Little, D. N. (1998). *A micromechanics fracture and healing model for asphalt concrete*. FHWA Final Report. DTFH61-92-C-00170, Texas A&M University.
28. Kim, Y. R., Little, D. N., & Benson, F. C. (1990). Chemical and mechanical evaluation on healing mechanism of asphalt concrete (with discussion). *Journal of the Association of Asphalt Paving Technologists, 59*, 240–275.
29. Little, D. N., Lytton, R. L., Williams, D. A., & Kim, Y. R. (2001). Microdamage healing in asphalt and asphalt concrete. In *Microdamage and microdamage healing, project summary report, FHWA-RD98-141, June 2001* (Vol. 1, p. 88).
30. Good, R. J., Manoj, K., & Chaudhuryvan Oss, C. J. (1991). Theory of adhesive forces across interfaces 2. In L.-H. Lee (Eds.), *Interfacial hydrogen bonds as acid—Base phenomena and as factors enhancing adhesion, Fundamentals of adhesion*. Springer Science Business Media, LLC.
31. Bhasin, A., Little, D. N., Bommavaram, R., & Vasconcelos, K. L. (2008). A framework to quantify the effect of healing in bituminous materials using material properties. *International Journal of Road Materials and Pavement Design, EATA, 08*(2008), 219–242.
32. FHWA. (2008). ARC consortium quarterly technical progress report. In *QuarterlyReport_April–June*. FHWA.
33. Qiu, J. (2012). *Self-healing of asphalt mixtures*. Technical University Delft.
34. Nahar, S. N., et al. (2014). Microstructural changes in bitumen at the onset of crack formation. *European Polymer Journal, 56*, 17–25.
35. Santagata, E., Baglieri, O., Dalmazzo, D., & Tsantilis, L. (2009). Rheological and chemical investigation on the damage and healing properties of bituminous binders. *Journal of the Association of Asphalt Paving Technologists, 28*(2009), 567–595.
36. Sun, D., Fan, Y., Li, L., Lin, T., & Zhu, X. Y. (2017). Effect of chemical composition and structure of asphalt binders on self-healing. *Construction and Building Materials, 133*, 495–501.
37. Hou, Y., Wang, L., Pauli, T., & Sun, W. (2015). Investigation of the asphalt self-healing mechanism using a phase-field model. *Journal of Materials in Civil Engineering, 27*(3).
38. Nahar, S. N. (2016). *Phase-separation characteristics of bitumen and their relation to damage-healing*.
39. Nazzal, M., Kaya, S., & Abu-Qtaish, L. (2012). Evaluation of the healing properties of WMA using atomic force microscopy. In *Proceedings of the 7th RILEM International Conference on Cracking Pavements* (Vol. 4). Springer.
40. Song, I., Little, D. N., Masad, E. A., & Lytton, R. (2005). Comprehensive evaluation of damage in asphalt mastics using X-ray CT, continuum mechanics, and micromechanics (with discussion). *Association of Asphalt Paving Technologists, 74*.
41. Grossegger, D. (2021). Fatigue damage self-healing analysis and the occurrence of an optimal self-healing time in asphalt concrete. *Journal of Materials in Civil Engineering, 33*(6), 04021098.
42. Gómez-Meijide, B., Ajam, H., Lastra-González, P., & Garcia, A. (2016). Effect of air voids content on asphalt self-healing via induction and infrared heating. *Construction and Building Materials, 126*(2016), 957–966.
43. GrosseggerGarcia, D. (2019). Influence of the thermal expansion of bitumen on asphalt self-healing. *Applied Thermal Engineering, 156*, 23–33.
44. Salih, S., Gómez-Meijide, B., Aboufoul, M., & Garcia, A. (2018). Effect of porosity on infrared healing of fatigue damage in asphalt. *Construction and Building Materials, 167*, 716–725.
45. Little, D. N., Prapnnachari, S., Letton, A., & Kim, Y. R. (1993). *Investigation of the micro-structural mechanism of relaxation and fracture healing in asphalt*. Texas Transportation Inst College Station
46. Wool, R. P., & O'Connor, K. M. (1981). A theory of crack healing in polymers. *Journal of Applied Physics, 52*(10), 5953–5963.

47. Wu, D.-Y., Meure, S., & Solomon, D. (2008). Self-healing polymeric materials: A review of recent developments. *Progress in Polymer Science, 33*, 479–522.
48. Kim, Y. H., & Wool, R. P. (1983). A theory of healing at a polymer-polymer interface. *Macromolecules, 16*(7), 1115–1120.
49. Tabaković, A., Schuyffel, L., Karač, A., & Schlangen, E. (2017). An evaluation of the efficiency of compartmented alginate fibres encapsulating a rejuvenator as an asphalt pavement healing system. *MDPI Applied Sciences, 7*, 16.
50. Moreno-Navarro, F., Sol-Sánchez, M., Rubio-Gámez, M. C. (2015). Exploring the recovery of fatigue damage in bituminous mixtures: The role of healing. *Road Materials and Pavement.*
51. Phillips, M. C. (1999). Multi-step models for fatigue and healing, and binder properties involved in healing. In *Eurobitume workshop on performance related properties for bituminous binders*, Luxembourg.
52. Lu, X., Soenen, H., & Redelius, P. (2003). Fatigue and healing characteristics of bitumens studied using dynamic rheometer. In *6th RILEM Symposium PTEBM'03, Zurich* (pp. 408–415).
53. Bhasin, A., Pavaldi, S., & Little, D. N. (2011). Influence of aging and temperature on intrinsic healing of asphalt binders. *Transport Research Board, 2207*, 9.
54. Hammoum, F., de La Roche, C., Piau, J.-M., & Stefani, C. (2002). Experimental investigation of fracture and healing of bitumen at pseudo-contact of two aggregates. In *Ninth ISAP International Conference on Asphalt Pavements, Denmark.*
55. Tozzi, C., Hammoum, F., & Baglieri, O. (2023). *Evaluation and modelling of low temperature fracture properties of asphalt binders in thin film by means of the local fracture test, road materials and pavement design* (pp. 1–17).
56. Maillard, S. (2005). *Etude du comportement à la rupture sous chargements répétés du bitume en film mince*, PhD report in French, Ecole centrale de Nantes.
57. Hyun-Jong L., & Kim Y. R. (1998). Viscoelastic continuum damage model of asphalt concrete with healing. *Journal of Engineering Mechanics, 124*(11), 1224–1232.
58. Erkens, S., van Vliet, D., & Leegwater, G. (2012). *Healing van asfaltmengsels, onderzoek naar een pragmatische proefmethode.* InfraQuest.
59. Pronk, A. C., & Cocurullo, A. (2009). Investigation of the PH model as a prediction tool in fatigue bending tests with rest periods. In *Advanced testing and characterisation of bituminous materials, Loizos, PartI.* Taylor & Francis Group.
60. Baaj, H., Mikhailenko, P., Almutairi, H., & Di Benedetto, H. (2018). Recovery of asphalt mixture stiffness during fatigue loading rest periods. *Construction and Building Materials, 158*, 591–600.
61. Di Benedetto, H., Ashayer, M., Soltani, A., & Chaverot, P. (1996). Fatigue damage for bituminous mixtures: A pertinent approach. *Journal of the Association of Asphalt Paving Technologists, 65*, 142–158.
62. Di Benedetto, H., de la Roche, C., Baaj, H., & Pronk, A. (1994). Fatigue of bituminous mixtures. *Materials and structures, 37*, 202–216.
63. Lundström, R., Ekblad, J., & Isacsson, U. (2004). Influence of hysteretic heating on asphalt fatigue characterization. *Journal of Testing and Evaluation, 32*, 484–493.
64. Redelius, P., & Soenen, H. (2015). Relation between bitumen chemistry and performance. *Fuel, 140*, 34–43.
65. Petersen, J. C., Robertson, R.-E., Branthaver, J. F., Harnsberger, P. M., Duvall, J. J., Kim, S. S., Anderson, D. A., Christiansen, D. W., & Bahia, H. U. (1994). *Binder characterization and evaluation: Volume 1, Rep. No. SHRP-A-367, Strategic Highway Research Program.* National Research Council.
66. Mewis, J., & Wagner, N. J. (2009). Thixotropy. *Advances in Colloid and Interface Science, 147–148*, 214–227.
67. Gauthier, G., Bodin, D., Chailleux, E., & Gallet, T. (2010). Nonlinearity in bituminous materials during cyclic tests. *Road Materials and Pavement Design, 11*(Supp 1), 129–158.
68. Mouillet, V., de la Roche, C., Chailleux, E., & Coussot, P. (2012). Thixotropic behavior of paving-grade bitumens under dynamic shear. *Journal of Materials in Civil Engineering, 25*, 23–31.

69. Santagata, E., Baglieri, O., Tsantilis, L., & Dalmazzo, D. (2013). Evaluation of self-healing properties of bituminous binders taking into account steric hardening effects. *Construction and Building Materials, 41*, 60–67.
70. Shan, L. Y., Tan, Y. Q., Underwood, B. S., & Kim, Y.-R. (2011). Separation of thixotropy from fatigue process of asphalt binder. *Transportation Research Record: Journal of the Transportation Research Board, 2207*, 89–98.
71. Delaporte, B., Van Rompu, J., Di Benedetto, H., Chaverot, P., & Gauthier, G. (2008). New procedure to evaluate fatigue of bituminous mastics using an annular shear rheometer prototype. In I. L. Al-Qadi, T. Scarpas, & A. Loizos (Eds.), *Pavement cracking: Mechanisms, modeling, detection, testing and case histories: Proceedings of the 6th RILEM International Conference on Cracking in Pavements* (pp. 457–467). CRC Press/Balkema
72. Di Benedetto, H., Nguyen, Q. T., & Sauzéat, C. (2011). Nonlinearity, heating, fatigue and thixotropy during cycling loading of asphalt mixtures. *Road Materials and Pavement Design, 5*, 163–202.
73. Soltani, A., & Anderson, D. A. (2005). New test protocol to measure fatigue damage in asphalt mixtures. *Road Materials and Pavement Design, 6*, 485–514.
74. Nguyen, Q. T., Di Benedetto, H., & Sauzéat, C. (2012). Determination of thermal properties of asphalt mixtures as another output from cyclic tension-compression test. *Road Materials and Pavement Design, 13*, 85–103.
75. Mangiafico, S. (2014). *Linear viscoelastic properties and fatigue of bituminous mixtures produced with Reclaimed Asphalt Pavement and corresponding binder blends.* PhD report, ENTPE.
76. Mangiafico, S., Sauzéat, C., Di Benedetto, H., Pouget, S., Olard, F., & Planque, L. (2015). Quantification of biasing effects during fatigue tests on asphalt mixes: non-linearity, self-heating and thixotropy. *Road Materials and Pavement Design, 16*, 73–99
77. Nguyen, Q. T. (2011). *Comportement thermomécanique des enrobés bitumineux sous sollicitations cycliques dans les domaines linéaire et non-linéaire.* PhD report, ENTPE.
78. Baaj, H. (2002). *Comportement à la fatigue des matériaux granulaires traités aux liens hydrocarbonés.* Dissertation, Lyon, INSA.
79. Baaj, H., Di Benedetto, H., & Chaverot, P. (2005). Effect of binder characteristics on fatigue of asphalt pavement using an intrinsic damage approach. *Road Materials and Pavement Design, 6*, 147–174.
80. Di Benedetto, H., et al. (2004). Fatigue of bituminous mixtures. *Materials and Structures, 37*, 202–216.
81. Tapsoba, N., Sauzéat, C., & Di Benedetto, H. (2013). Analysis of fatigue test for bituminous mixtures. *Journal of Materials in Civil Engineering, 25*, 701–710.
82. Williams, D. et al. (2001). *Micro-damage healing in asphalt and asphalt concrete, volume II: Laboratory and field testing to assess and evaluate micro-damage and micro-damage healing.* No. FHWA-RD-98-14.
83. Sun, D., Yu, F., Li, L., Lin, T., & Zhu, X. Y. (2017). Effect of chemical composition and structure of asphalt binders on self-healing. *Construction Building Materials, 133*, 495–501.
84. Little, D. N., Bhasin, A., & Darabi, M. K. (2015). Damage healing in asphalt pavements: Theory, mechanisms, measurement, and modeling. In *Advances in asphalt materials* (pp. 205–242). Woodhead Publishing.
85. Nahar, S., Schmets, A., Kasbergen, C., Schitter, G., & Scarpas, A. (2016). *Microstructural changes in bitumen at the onset of damage-healing.* CROW InfraDagen.
86. Bhasin, A., Palvadi, S., & Little, D. N. (2011). Influence of aging and temperature on intrinsic healing of asphalt binders. *Transportation Research Record: Journal of the Transportation Research Board, 2207*, 70–78.
87. Kausch, H. H. (1987), Fracture mechanics studies of crack healing. In *Polymer fracture* (2nd ed.).
88. Boiko, Y. M., & Lyngaar-Jorgensen, J. (2004). Healing of interfaces of high and ultra-high molecular weight polystyrene below the bulk Tg. *Polymer, 45*(25), 8541–8549.

89. Kinloch, A.-J. (1987). Adhesion and adhesives. In *Science and technology* (2nd ed.). Chapman and Hall.
90. Bhasin, A., Little, D. N., Bommavaram, R., & Vasconcelos, K. (2008). A framework to quantify the effect of healing in bituminous materials using material properties. *Road Materials and Pavement Design, 9*, 219–242.
91. Hou, Y., Wang, L., Pauli, T., & Sun, W. (2015). Investigation of the asphalt self-healing mechanism using a phase-field model. *Journal of Materials in Civil Engineering, 27*, 04014118.
92. Leegwater, G., Scarpas, T., & Erkens, S. (2016). Direct tensile test to assess healing in asphalt. *Transportation Research Record: Journal of the Transportation Research Board* 124–30.
93. Little, D. N., Bhasin, A., & Darabi, M. K. (2015). Damage healing in asphalt pavements: Theory, mechanisms, measurement, and modeling. Theory, mechanisms, measurement, and modeling. *Advances in Asphalt Materials: Road and Pavement Construction.*
94. Nazzal, M., Kaya, S., & Abu-Qataish (2012). Evaluation of WMA healing properties using atomic force microscopy. In T. A. Scrpas (Ed.), *7th Rilem Conference on Advanced Testing and Characterization of Bituminous Materials* (pp. 1125–1134).
95. Santagata, E., Baglieri, O., Dalmazzo, D., & Tsantilis, L. (2009). Rheological and chemical investigation on the damage and healing properties of bituminous binders. *Asphalt Paving Technology-Proceedings, 28*, 567.
96. Soenen, H., Besamusca, J., Fischer, H. R., Poulikakos, L. D., Planche, J.-P., Das, P. K., Kringos, N., Grenfell, J. R. A., Lu, X., & Chailleux, E. (2013). Laboratory investigation of bitumen based on round robin DSC and AFM tests. *Materials and Structures* 1–16.
97. Song, I., Little, D., Massad, E., & Lytton, R. L. (2005). Comprehensive evaluation of damage in asphalt mastics using X-ray CT, continuum mechanics, and micromechanics. *Journal of Assoc Asphalt Paving Technology, 74*, 36.

Extrinsic Self-Healing Technology for Asphalt Pavements

Amir Tabaković, Greet Leegwater, Quantao Liu, Jose Norambuena-Contreras, Hassan Baaj, and Orazio Baglieri

Abstract Extrinsic self-healing technology is a new field within road construction materials. Incorporating extrinsic self-healing technology into the road design process can transform construction and maintenance by increasing life spans and reducing maintenance needs. By delaying the premature deterioration of asphalt pavements, extrinsic self-healing asphalt may reduce natural resources consumption, decrease the traffic disruption caused by road maintenance processes, decrease CO_2 emissions during the road maintenance process and increase road safety. In addition to environmental benefits, self-healing construction materials have the potential to significantly reduce costs associated with road network maintenance. Currently, five extrinsic self-healing technologies are available for asphalt pavement design: nanoparticles, induction heating, microwave heating, rejuvenator encapsulation and hybrid (combination of induction and rejuvenator encapsulation). This

A. Tabaković (✉)
Revolution Road Ltd, Bray, Wicklow, Ireland
e-mail: amir.tabakovic@revolutionroad.eu

School of Civil Engineering, University College Dublin, Belfield, Ireland

G. Leegwater
Netherlands Organisation for Applied Scientific Research (TNO), NEXT Delft, Delft, The Netherlands

Q. Liu
State Key Laboratory of Silicate Materials for Architectures, Wuhan University of Technology, Wuhan, China

J. Norambuena-Contreras
Department of Civil Engineering, Faculty of Science and Engineering, Materials and Manufacturing Research Institute, Swansea University, Bay Campus, Skewen, UK

H. Baaj
Department of Civil and Environmental Engineering, University of Waterloo, 200 University Ave W, N2L 3G1, Waterloo, ON, Canada

O. Baglieri
Department of Environment, Land and Infrastructure Engineering, Politecnico di Torino, 24, c.so Duca degli Abruzzi, 10129 Turin, Italy

© The Author(s), under exclusive license to Springer Nature Switzerland AG 2025
H. Baaj and O. Baglieri (eds.), *Crack-Healing of Asphalt Pavement Materials*, RILEM State-of-the-Art Reports 45, https://doi.org/10.1007/978-3-031-99443-2_3

chapter reviews all five options and outlines the future development of extrinsic self-healing asphalt technology.

1 Examples of Extrinsic Self-Healing Technology for Asphalt Pavements

The self-healing performance of an asphalt pavement can be improved through modifiers and additives introduced to the asphalt mix targeting its self-healing properties. To work in an asphalt pavement system, the self-healing technology needs to survive the harsh conditions prevalent during asphalt pavement construction (i.e., mixing and compaction) and service life (i.e., traffic loading and environmental conditions, such as rain, ice, snow, and high temperatures). Qiu et al. in 2009 [1] outlined five essential conditions for extrinsic self-healing to be included in asphalt pavement design:

1. good compatibility with asphalt binder,
2. high-temperature stability,
3. the ability to survive mixing and construction conditions,
4. a healing temperature of between − 30 and 40 °C and
5. capable of continuous/multi-time healing.

In this section, several existing extrinsic self-healing technologies used to aid asphalt's natural self-healing abilities are presented and critically analysed with respect to their functional design and performance. Self-healing methods found to be available during the lifetime of TC 278-CHA can be classified into five categories:

1. Nanoparticle-based modifiers,
2. Induction heating,
3. Microwave heating,
4. Encapsulated rejuvenators and
5. Hybrid—combination of induction heating and encapsulated rejuvenators.

2 Nanoparticle-Based Modifiers

2.1 *Carbon Nanotubes*

Due to their unique mechanical properties and their large surface area to volume ratio, the use of carbon nanotubes as potential modifiers of asphalt materials have attracted the interest of researchers in recent years [2]. It has been demonstrated that carbon nanofibres can improve the fatigue resistance of asphalt materials by means of crack bridging and pull-out mechanisms [3]. Actual reinforcement effect of carbon nanotubes is dependent on the physico-chemical properties of base components, which influence the morphological configuration assumed by particles within the asphalt matrix. The efficacy of carbon nanotubes is also influenced by degree of dispersion attained during mixing operation; the use of high shear mixing combined with ultrasounds has been proven to be an effective technique to prepare asphalt materials modified with carbon nanotubes [4].

Rod bridging provided by carbon nanotubes can contribute to controlling the spacing between crack fronts, thus triggering the self-healing phenomena. Moreover, even if the higher viscosity of binders containing carbon nanotubes can delay the flow phase leading to wetting, this negative effect can be offset by the aptitude of nano-particles to alter the morphology of crack interfaces [5–8]. In fact, more fractal surfaces generated by the presence of carbon particles in the matrix can promote interfacial interactions that represent the driving force of the wetting phase of the healing process.

2.2 *Nanoclay*

Nano-sized clay is introduced into the asphalt mixture as a modifier to improve the ageing, rheological and thermal properties of asphalt composite [9]. The nanoclay particles also have the potential to repair micro-cracks in asphalt binder. In fact, nano-sized particles tend to move towards the crack's tip, driven by surface energy; at the crack tip, they are expected to stop crack propagation and heal damaged asphalt material [1].

Liu et al. [10] and Wu et al. [11] indicated that an adequate interfacial interaction between asphalt binder and nanoclays particles can be beneficial in improving fatigue and healing properties of asphalt binders. Tabatabaee and Shafiee [12] studied the effect of rest periods on the fatigue life of organoclay-modified asphalt mixes. They found that introducing rest periods at 3 and 5% strain level increased the fatigue resistance of asphalt mixes containing organoclay-modified asphalt binders.

It has been postulated that the morphological configuration assumed by clay sheets is a key factor in the definition of properties of the composite material [13], the performance of which may be linked to the obtained degree of intercalation and/or exfoliation. This can be attributed to the aptitude of alkyl chains, which constitute the tails of the nanoclay surfactant, to promote the diffusion and randomisation phases of the healing process through reptation-type interactions [14]. Hence, nanoclay characterised by a more exfoliated morphology is believed to be able to activate a larger number of surfactant molecule chains with positive influence on healing mechanisms [15]. Such findings illustrate that nanoclay material can potentially improve the healing capability of asphalt mixtures.

However, the nanoclay self-healing technique has not been researched in-depth to date. Very little is known about the long-term effect of the nanoclay particles on the self-healing asphalt mix performance. Therefore, future studies should focus on the impact of temperature and time on the asphalt pavement healing performance.

2.3 Nanorubber

As with nanoclay materials, polymer and rubber modifiers are used in the asphalt materials to improve the binders' physical and mechanical properties and the in-situ performance of asphalt pavements [16–18]. Rubber modifiers in the form of nanoparticles have also been used to enhance the healing properties of asphalt mastic [9]. Qiu et al. [1] studied the repair of asphalt mastic with a nanorubber-modified binder (70/100 pen binder). They conducted ductility recovery tests to assess the self-healing capacity of the asphalt mastic, containing two different types of nanorubber modifier (NanoA and NanoB) and varying percentages of the modifier (0–5%), where 0% was a control mix. The tests were performed on tensile test specimens. The test specimens were cut at the centre, joined and left to heal for 4 h at room temperature (20–22 °C). The test results showed good self-healing for non-modified asphalt mastic mixes, up to 70%.

Recovery of the modified asphalt mixtures varied from 15 to 90% depending on the type of nanorubber modifier and the amount. Unfortunately, the authors did not present the composition of the nanorubber modifier, so it is difficult to determine what type of healing mechanism activated the healing process and to understand the variability of the healing performance. Nevertheless, the study shows that some nanorubber modifiers can be used to improve the self-healing properties of asphalt mixtures. Future studies should show the properties of nanorubber modifiers, i.e., whether they are styrene-butadiene-rubber (SBR), styrene–isoprene-styrene (SIS) or olefinic rubber. This would determine what asphalt healing mechanism has taken place, such as: reverse hydrogen bonding. The clear advantage of this self-healing modifier is its double role; it can not only improve asphalt mix durability but also be used as a self-healing modifier in the mix.

The disadvantage of polymer-based modifiers is the thermodynamic incompatibility with asphalt binder due to large differences in density, polarity, molecular weight and solubility between the polymer and the asphalt. This can result in delamination of the composite during thermal storage, which is not readily apparent and adversely affects the asphalt mix when used [9]. The thermodynamic incompatibility of nano modifiers requires investigation, as high temperatures may cause separation between the nanorubber modifier and binder, potentially degrading asphalt pavement performance.

The nano effect, along with reverse hydrogen and ionic bonding, are recognized for their multi-healing capabilities [1], which is a benefit of this self-healing mechanism. However, the effects of time and temperature on the healing aspect of the asphalt mix need to be investigated. If ineffective at lower temperatures and without improvements in healing times relative to intrinsic asphalt binder healing, the technology will be rendered unsuitable for the self-healing of asphalt pavements. Although the nanoparticle self-healing technology demonstrated its potential in asphalt pavement mix design, more substantial evidence of its performance needs to be illustrated before it becomes accepted as a viable self-healing technology.

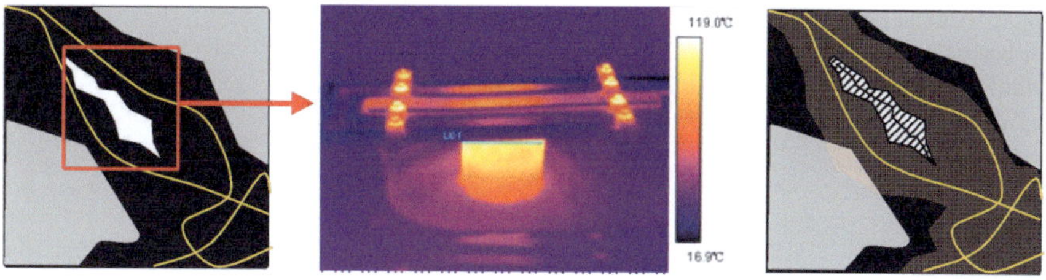

Fig. 1 Diagram of induction healing in porous asphalt (Reproduced from Liu [19])

3 Induction Heating

3.1 Introduction

The second technology available to create extrinsic healing is induction heating. This concept involves adding electrically conductive and/or magnetic materials (e.g., metallic particles or fibres) into the asphalt mixture. An alternating electromagnetic field can then be generated using an induction heating device to heat the particles or fibres. In this process, heat is generated by the eddy currents occurring within the electrically conductive materials. Eddy currents create magnetic fields opposing the induced field, causing Joule heating. Finally, through heat conduction, the asphalt binder around these conductive additives will be heated, decreasing its viscosity, eventually flowing and closing cracks nearby, and recovering the adhesive bonds with the mineral aggregates. This method can be repeated if damage reoccurs. A schematic diagram of induction healing is illustrated in Fig. 1.

The major healing mechanism of induction healing is the capillary flow caused by the thermal expansion of asphalt binder. Garcia [20] verified this healing mechanism using the asphalt binder capillary flow combined with tomography tests.

3.2 Self-Healing Capacity of Asphalt Concrete with Induction Heating

The effect of induction heating cycles on the performance of asphalt, including steel fibres, has been investigated using different approaches. A selection will be presented in this paragraph. Liu and Schlangen [21] studied the induction healing effect of steel fibres and steel wool, characterizing asphalt healing using the following equation:

$$HI = \frac{C_2}{C_1} HI = \frac{C_2}{C_1} \tag{1}$$

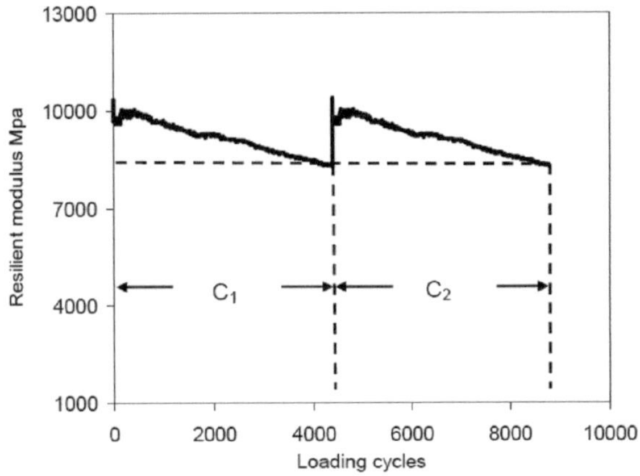

Fig. 2 Fatigue recovery of porous asphalt mixture cylinder for resilient modulus reduction to 80% (Reproduced from Liu [19])

where HI is the healing index (%), where 100% is the complete healing of damage and 0% is no healing at all; C_1 is the number of the loading cycles for the first-time loading, and C_2 is the number of loading cycles for the second time loading (after healing).

Figure 2 presents a fatigue test result by Liu [19] showing two values of the loading cycle C_1 and C_2. The loading cycle C_2 illustrates the effect of induction healing. If the fatigue tests were stopped when the resilient modulus reduced to 80% of its initial value, the fatigue damage could be completely healed.

Additionally, Liu and Schlangen [22] investigated the effect of long steel wool (approximately 10 mm) versus short steel fibre (approximately 0.2 mm). The results showed that long steel wool is more effective than short steel fibre in making electrically conductive porous asphalt concrete. They also demonstrated that the addition of steel fibres affects the mechanical performance of asphalt by reinforcing the mastic (asphalt binder, filler and sand) of porous asphalt mixture, which delays the ravelling effect in asphalt pavements. Liu and Schlangen [21] also demonstrated that the inclusion of steel fibres into the asphalt pavement mix prevents the drainage of asphalt binder from the surface of the asphalt pavement. The advantage of adding fibres is that a better bond between the large aggregates (stones) in the pavement is achieved.

Liu et al. [22] evaluated the healing capacity of a porous asphalt mixture containing steel wool fibres under induction heating through a four-point bending fatigue test. The results shown in Fig. 3 demonstrated that the fatigue life of porous asphalt mixture could be significantly extended by applying induction heating due to a faster flow of asphalt binder. The optimal induction heating temperature for the porous asphalt mixture was 85 °C, and the maximum fatigue life extension ratio reached 62.7%. They also investigated the possibility of multiple instances of induction heating and showed that induction heating could be repeated when cracks return after healing. With four instances of damage loading of 50,000 cycles followed by four instances of induction heating and resting, the modified fatigue life is 277,720 cycles, 2.9 times the original fatigue life. This represents a fatigue life extension ratio of 190% (see Fig. 4). Even though this research does not fully optimise when to heat the pavement and heating frequency, multiple instances of heating can significantly extend the fatigue life of the asphalt mixture.

Dai et al. [23] also evaluated the healing capacity of electroactive asphalt mastic and concrete beam samples with induction heating. They also used electrically

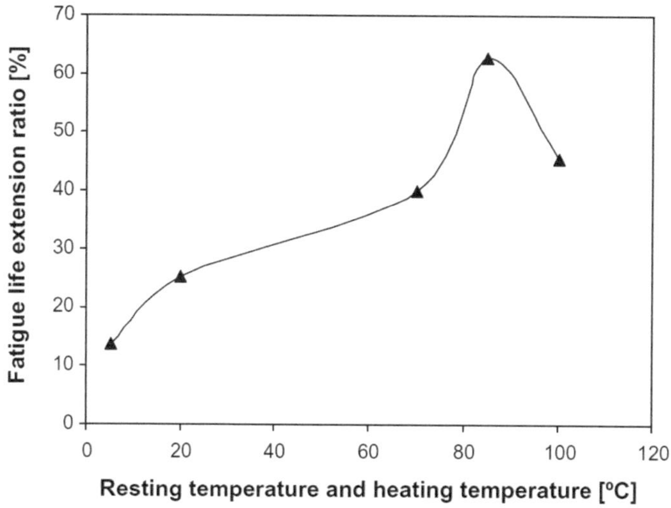

Fig. 3 Fatigue life extension ratios (at 300 microstrains) of the samples at different heating temperatures (Reproduced from Liu et al. [22])

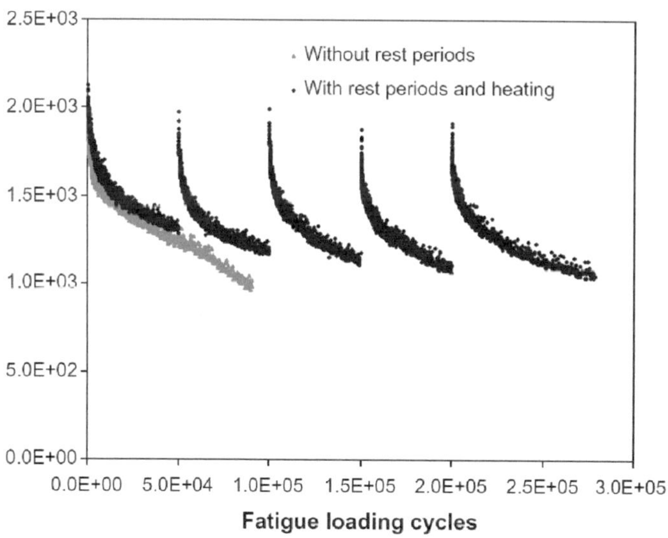

Fig. 4 Fatigue life extension caused by multiple induction heating (Reproduced from Liu et al. [22])

conductive steel wool fibres. They showed that the healing performance increased with the heating temperatures and that part of the original fracture strength was maintained after several healing cycles.

However, García and Schlangen [24] reported the existence of a limiting volume fraction of fibres at which all the fibres were surrounded by asphalt binder. Above this value, the temperature does not increase if additional fibres are added to the mix since energy cannot be transmitted to the asphalt. Furthermore, this value coincided with the quantity needed to reach the maximum conductivity of the mastic. Below this optimum volume fraction, the mastic behaves as a non-conductive material, although the heating effect still exists due to local conductivity [22]. To determine the optimum fiber content, García and Schlangen [25] applied percolation theory. They progressively increased the volume of conductive additives and analyzed each mixture separately until a sudden drop in resistivity was observed. It was proven that induction heating did not damage the mastic, and the number of times a test specimen could be healed depended on the maximum temperature reached. Furthermore, the sand–asphalt binder ratio was a key factor in the design of conductive asphalt mixtures and could not be considered separately from the amount of conductive particles in the asphalt mix. To increase the sand-asphalt binder ratio, they added graphite as an electromagnetic conductive additive.

Researchers have demonstrated that asphalt binder properties are a key factor influencing the control of the asphalt healing process. It has been observed that the asphalt binder must be able to flow as a near-Newtonian fluid to close the cracks [26] efficiently. For this reason, the temperature of the asphalt material must be increased above a certain limit (heating phase) and the material needs to be continuously heated for some time afterward. The healing temperature depends on the type of asphalt binder and its rheological properties. This aligns with observations on intrinsic healing, which show more recovery in the case of a high percentage of asphalt binder or a soft binder.

Liu [19] demonstrated that the degree of damage also affects the healing ratio, indicating that heating should be applied precisely for optimal results. If applied too early (when the resilient modulus is > 80% of the original asphalt pavement stiffness value), the asphalt pavement can heal through intrinsic healing and extrinsic healing is not required. However, if the asphalt is heated too late (when the resilient modulus value is < 70% of the original asphalt pavement stiffness value), the healing effect will be poor because structural damage, such as permanent deformation or stone aggregate cracking, can occur. Such damage is beyond the healing capacity of the induction healing process. Liu further concluded that the best healing results were achieved when induction heating was combined with a rest period. In Liu's study, he showed that complete healing could be achieved if the intrinsic healing process is aided by extrinsic healing through induction heating.

Although it has been demonstrated that induction heating can enhance the self-healing capacity of asphalt pavements, an adverse effect of the approach is that heating the asphalt mix mat would age the asphalt binder, especially when induction heating is repeated [27]. The effective healing depth of the asphalt mixture is limited around 43 mm due to the gradient heating phenomenon of asphalt concrete under induction heating [28]. Furthermore, overheating (> 100 °C) the asphalt mix can cause binder swelling and drainage, which will adversely affect the pavement performance. Therefore, Liu and Schlangen [22] suggested that the optimal heating temperature for the porous asphalt mix made with pen 70/100 asphalt binder is 85 °C.

In December 2010, researchers from TU Delft, in cooperation with the Dutch National Roads Authority, resurfaced a road (the A58 in the Netherlands) using an induction self-healing asphalt mix containing 10 mm long steel wool fibres [27]. The testing of the self-healing process within the Netherlands A58 motorway is ongoing at present. Induction heating equipment for efficient field heating needs to be developed to apply this healing technology.

4 Microwave Heating

4.1 Introduction

The third extrinsic healing approach is microwave heating. In microwave heating, an electromagnetic radiation device is used to subject asphalt mixtures to an alternating electromagnetic field with a frequency in the order of Megahertz [29, 30]. Microwave heating affects the asphalt binder by changing the orientation of polar molecules, which results in internal friction and an increase in the mixture temperature [31]. So, when ferrous materials are added to the mixtures, they can reflect microwave radiation and accelerate the temperature-increasing process. Hence, steel wool fibres, conductive filler or nanoparticle additives/modifiers can increase the heating rates of asphalt mixtures since they can absorb more thermal energy than the aggregates and asphalt binder. A schematic diagram of the microwave healing process of asphalt mixtures with steel wool fibres is illustrated in Fig. 5.

Although the scientific contributions in induction and microwave heating applications have been diverse, there is currently no criterion for selecting the most appropriate technology for the electromagnetic heating of asphalt materials. In conventional heating processes on asphalt mixtures, e.g., infrared heating and oven heating, the energy is transferred to the asphalt material by convection, conduction, and radiation of the heat from the surfaces of the material. However, in other processes, such

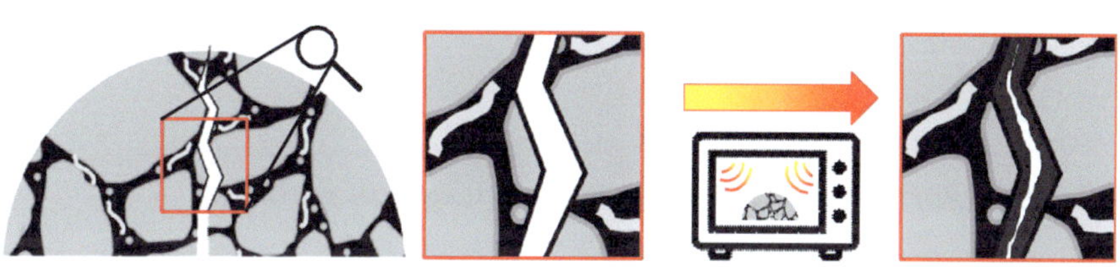

Fig. 5 Microwave healing of asphalt mixture (Reproduced from Norambuena-Contreras and Gonzalez-Torre [32])

as microwave heating, heat is directly delivered to the materials through molecular excitation [33]. This difference can result in many potential advantages of using microwaves for processing materials, especially for relatively thick materials [34].

Gallego and del Val [30] studied microwave heating as a potential extrinsic healing method for asphalt self-healing. They found that it is possible to employ microwaves to heat asphalt mixtures and use this technique to achieve self-healing of asphalts by heating them during rest periods. Steel wool fibres were added into asphalt mixtures to enhance the microwave heating effect. Their study shows that in microwave heating, the steel wool fibre content in the asphalt can be lower than the recommended induction heating content. This can reduce material costs.

Norambuena-Contreras and Gonzalez-Torre [32] studied the influence of microwave heating time on the self-healing properties of asphalt mixtures. In this research, the healing properties of asphalt mixtures with steel wool fibres were measured using the 3-point bending test before and after microwave healing at four different heating times, from 10 to 40 s, and during ten cracking-healing cycles. The key finding of this study was that for microwave heating, the heating time is the most influential variable of the healing level. Also, CT-Scan results proved that steel wool fibre spatial distribution within the asphalt mixtures plays an important role in the asphalt healing level. Besides, it was concluded that the optimum microwave heating time is 40 s to reach the highest healing levels with the lowest damage to the asphalt.

In this context, González, and Norambuena-Contreras [35] evaluated the crack-healing performance of asphalt mixtures containing varying percentages of reclaimed asphalt pavements (RAP) and metallic waste fibres. The study showed that the addition of metallic fibres into the asphalt mix increases the crack-healing performance of the mixtures, while RAP decreases the healing performance of the mixture. A disadvantage of adding steel wool fibres into an asphalt mix is that fibres tend to cluster together and thus increase the air void content in the mixture and uneven distribution of heat throughout the asphalt pavement. The increase in porosity could potentially reduce the physical and mechanical properties of the asphalt mixture.

Since the aggregates of an asphalt mixture are also sensitive to heating by microwave radiation, Trigos et al. [36] studied the dielectric properties of different types of aggregates to evaluate their susceptibility to microwave heating. They found that aggregates with higher susceptibility are of igneous origin (volcanic and sub-volcanic), exhibiting very high to medium susceptibility, followed by plutonic, metamorphic (foliated and non-foliated), and sedimentary aggregates, with low to very low susceptibilities. These results could be used to optimize the composition of asphalt mixtures for microwave self-healing treatments.

4.2 Comparison of Microwave and Induction Heating

Norambuena-Contreras and Garcia [29] compared the healing effect of induction and microwave heating technologies for the first time; see Fig. 6. They reported that microwave heating only increased the temperature of the binder and that the

Fig. 6 Self-healing mechanism via **a** microwave and **b** induction heating (Reproduced from Norambuena-Contreras and Garcia) [29]

metallic fibres became very hot when exposed to microwave radiation. As a result, it is difficult to control the asphalt binder temperature during microwave heating. The microwaves heat the asphalt mixture to heal the cracks, but at the same time, they may degrade asphalt binder and change the air void structure inside the asphalt mixture, affecting the mechanical strength of the asphalt mixture. In short, as the temperature of the binder under microwave radiation can be higher than the flash point temperature of asphalt binder, microwave heating may modify the chemical structure of the binder. So, for future works using microwave radiation, authors recommend microwave heating times under 40 s.

5 Encapsulated Rejuvenator Techniques

5.1 Introduction

The addition of encapsulated rejuvenators into the asphalt mixtures to restore the original properties of aged asphalt binder is a promising self-healing method studied by many researchers [37–75]. Ageing in asphalt binder is caused by (1) evaporation of the volatile components, (2) oxidation, and (3) polymerisation [76]. As a result of ageing, the asphalt binder becomes stiffer and more brittle [77, 78]. Although the chemistry behind this process is irreversible, the original viscoelastic properties of the asphalt mixture can be recovered by adding softer asphalt binder or a rejuvenating agent such as cationic emulsions [79–81]. The problem is that this type of treatment is superficial: only the first centimeters from the surface are affected. To solve this problem, it was thought that the optimum way of adding maltenes to the road would be by mixing capsules of different sizes filled with maltenes with asphalt mixtures [38].

The work principle of the encapsulated rejuvenators can be resumed as follows. When a microcrack appears in the asphalt pavement and reaches a capsule, the fracture energy at the tip of the crack stresses the capsule and breaks it, releasing the

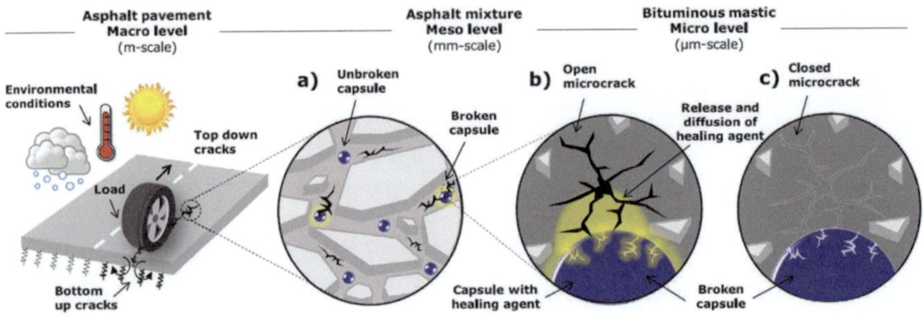

Fig. 7 Self-healing mechanism of encapsulated rejuvenators in asphalt mixture. **a** capsules with rejuvenator dispersed in the asphalt matrix, **b** broken capsules and rejuvenating agent diffusion process in the aged asphalt binder, and **c** rejuvenator restoring the original chemical balance of asphalt binder, sealing the open microcrack. (Reproduced from Gonzalez-Torre and Norambuena-Contreras [3])

rejuvenating agent. Then, the rejuvenator flows into the microcracks due to capillarity and diffusion processes, softening the surrounding asphalt binder, restoring the original chemical balance of the asphalt binder, and making it flow through the microcrack and seal it [37]. Figure 7 illustrates the healing process using encapsulated rejuvenators. The process prevents the propagation of micro-cracks within the asphalt pavement and will prevent the complete failure of the pavement system. Thus, several researchers demonstrated that various types of polymer-based capsules and fibres containing rejuvenators can be produced. These encapsulated rejuvenator types are thermally and mechanically stable to survive the asphalt production process.

One of the crucial elements to consider in the rejuvenation self-healing technique is to select an appropriate rejuvenator. The primary purpose of the rejuvenator in the self-healing system is to diffuse into the aged binder and recover its chemical structure, which was damaged during the pavement's service life because of ageing.

A rejuvenator is an engineered cationic emulsion containing maltenes and saturates. The primary aim of a rejuvenator is to reduce the stiffness of the oxidised asphalt binder to extend the pavement's service life [80].

Currently, many researchers have explored the use of alternative asphalt binder rejuvenator oils coming from natural resources, such as sunflower oil [45–63], bio-oil from liquefied biomass waste [64], and rapeseed oil [65], among others. Additionally, oils coming from man-made resources have also been proposed as rejuvenators, such as waste cooking oil [66–69], virgin [67] and waste [69] engine oil, and recently pyrolytic oil from end-of-life tyres [70]. Su et al. [66] showed that rejuvenators produced from recycled waste cooking oil (WCO) can be encapsulated and used in asphalt self-healing technology. The capsules can survive the harsh asphalt production process, with high capsule shell strength of 1.0–1.52 GPa—c/s ratio of 1:3 and good thermal stability, with a melting point of 180 °C. This research demonstrated that WCO rejuvenator can successfully penetrate and rejuvenate a standard asphalt binder of 80/100 pen value and diffuse it at a low temperature (0 °C), see

Table 1. The results illustrate the return of the aged binder to the same workability as the original binder.

Garcia et al. [47] used sunflower oil to repair the damage and recover aged binder properties. Tabaković et al. [54] studied the performance of different types of rejuvenators as healing agents in an asphalt mastic mixture. They reported that vegetable oil and waste cooking oil performed well, with specimens recovering 25% and 30% of their original tensile strength, respectively. However, the rejuvenator also led to a higher deformation in the asphalt test specimen, which on-site could result in another pavement damage mechanism known as rutting. They reported that the observed performance of an industrially designed rejuvenator was more favourable. It was able to recover 50% of the original test specimen strength after macro cracking, while also recovering its original material stiffness. As such, the authors decided to use industrially produced rejuvenators in the rest of their studies [42–44].

Concha and co-workers [67] compared the performance of sunflower oil (VCO), waste cooking oil (WCO) and virgin engine oil (VEO) as rejuvenators for PAV-aged asphalt binder. They proved that sunflower oil presented a better rejuvenating performance. The addition of 2.2% per volume of asphalt binder restored the saturates and aromatic fractions of PAV-aged asphalt binder to values close to a RTFO-aged asphalt binder. They also simulated the release of different amounts of rejuvenators from a capsule to close an artificial 150 micro-crack in a PAV-aged asphalt binder sample. It was found that the VCO presented the best performance inclosing the 150 micro-crack, where an amount of oil equivalent to the release of 50, 75 and 100% of the encapsulated oil closed the crack after 78, 35, and 20 min, respectively (see Fig. 8).

Norambuena-Contreras et al. [70] proposed the use of the pyrolytic oil coming from the pyrolysis of end-of-life-tyres as a potential rejuvenator for asphalt self-healing purposes. They showed that the pyrolytic oil was composed of 98.07% wt. maltenes, with a low proportion of asphaltenes (1.96% wt.) and components with antioxidant activity, making this oil suitable for rejuvenation. The study reported that the application of 1 mg of the oil was able to close an artificial 150 micro-crack on a short-term aged asphalt binder, proving that the pyrolytic oil is a rejuvenator that

Table 1 Properties of virgin and WCO rejuvenated asphalt binder [66]

Material property	Original asphalt binder (80/100 pen)	Aged asphalt binder (40/50 pen)	Rejuvenated aged asphalt binder with micro WCOs (wt.%)					
			2	4	6	8	10	12
Penetration @ 25 °C (d-mm)	86	43	52	61	74	82	84	88
Softening point (°C)	46.7	53.5	52.0	51.2	50.3	48.4	46.4	45.0
Viscosity @ 135 °C (mPa s)	325	578	570	520	470	390	333	310

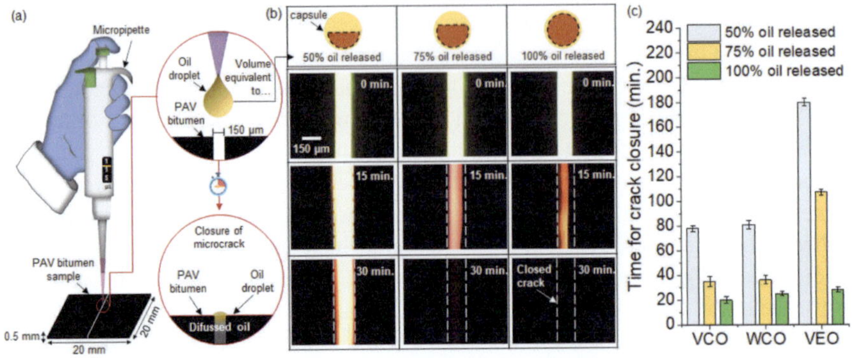

Fig. 8 **a** Crack closure healing test on an PAV aged asphalt binder sample; **b** Fluorescence microscopy images showing the crack closure over time using different amounts of sunflower oil as rejuvenator, and **c** times required for each type and amount of oil released from a capsule to close the microcrack (Reproduced from Concha et al. [67])

can diffuse in the aged asphalt binder, reducing its viscosity to promote the asphalt self-healing, see Fig. 9.

Another crucial element to consider in the encapsulated rejuvenator method is to select an appropriate shell to encapsulate the rejuvenator. Five types of encapsulating methods have been used so far: (1) Saturating porous sands with a rejuvenator and sealing them with an epoxy-cement matrix. (2) Microencapsulation of rejuvenator (core–shell) by in-situ polymerisation. (3) Multi-cavity calcium-alginate capsules, (4) Biobased spore microcapsules, and (5) Compartmented calcium-alginate-fibres.

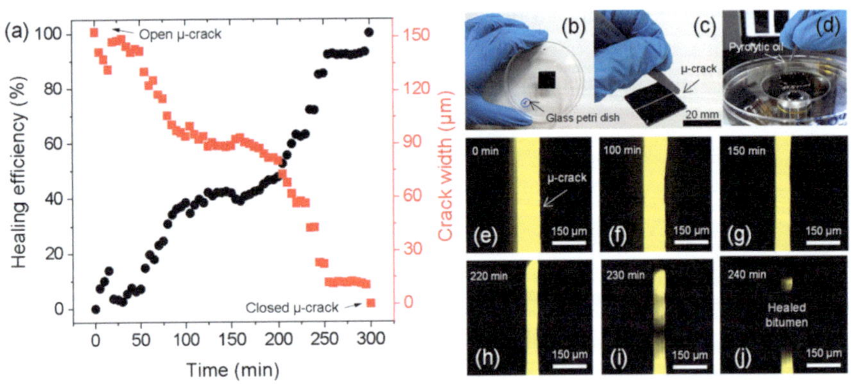

Fig. 9 **a** Curves of healing efficiency and crack width over time on a short-term aged asphalt binder by effect of the pyrolytic oil; **b–d** Representation of the healing test, and **c** fluorescence microscopy images showing the closure of the 150 micro-crack with time (Reproduced from Norambuena-Contreras et al. [70])

5.2 Epoxy-Based Macrocapsules

Garcia et al. [47, 48] successfully produced macrocapsules using porous sand as a rejuvenator (sunflower oil) absorbing material. The sand granules are bound together and coated by a hard shell made of an epoxy-cement matrix. The resulting capsules consisted of a varying percentage of rejuvenator (13, 21 and 25%), porous sand, cement, and epoxy. Figure 10 shows epoxy macrocapsule morphology.

The capsules' mean diameter was 1.60 mm. A part of the sand aggregates in the asphalt mixture was substituted with the capsules. The working principle of the epoxy capsule is as follows: when the stress in capsules embedded in the asphalt pavement reaches a certain threshold value, the capsules break, and the rejuvenator is released. This type of capsules could resist the mechanical forces during the mixing and compaction process of asphalt mixture and break under cyclic compression test. The asphalt mixture showed signs of crack closure and air void reduction after a resting period. However, these capsules decreased the stiffness, indirect tensile strength, and rutting resistance of the mixture due to the lower strength of the capsules and their breakage.

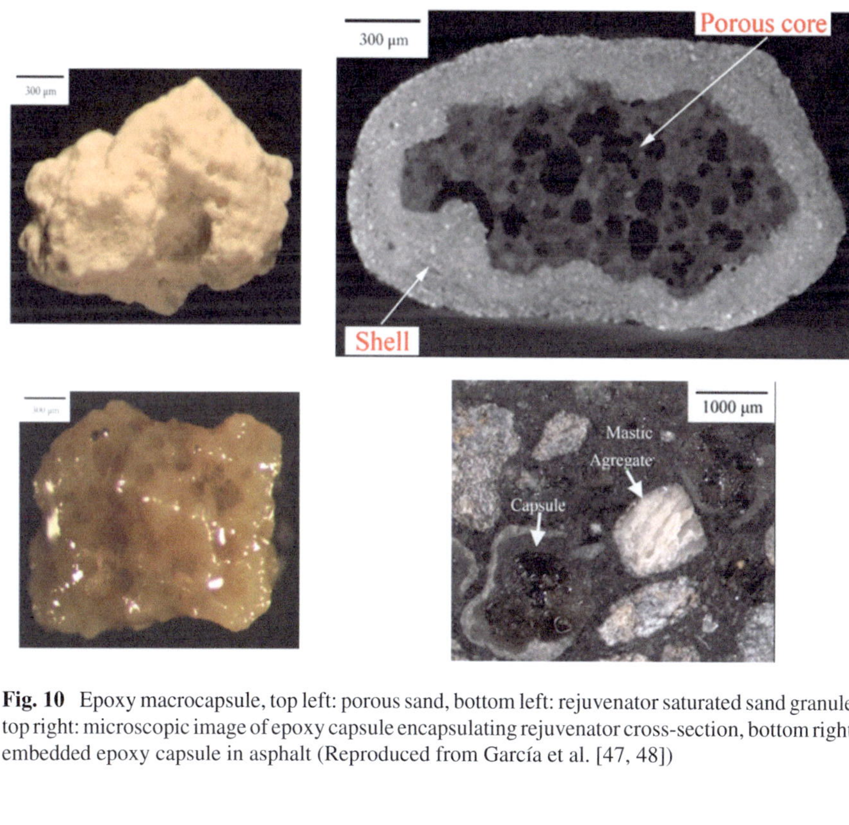

Fig. 10 Epoxy macrocapsule, top left: porous sand, bottom left: rejuvenator saturated sand granule, top right: microscopic image of epoxy capsule encapsulating rejuvenator cross-section, bottom right: embedded epoxy capsule in asphalt (Reproduced from García et al. [47, 48])

5.3 *Polymer-Based Microcapsules*

Another successful microcapsule with asphalt self-healing purposes is one where the shell material is made of a prepolymer melamine–formaldehyde modified by methanol (78.0% solid content), and the rejuvenator is an oily product [55]. Figure 11 illustrates the production process of the microcapsules containing rejuvenators by in-situ polymerisation. The efficiency of the microcapsule fabrication is measured by the amount of rejuvenator retained within the microcapsule [40]. The highest efficiency achieved is 70% [39] under the following production conditions: a core to shell ratio of 1:3, stirring speed of 3000 rpm and 2.0–2.5% Styrene maleic anhydride (SMA) copolymer dispersant. Figure 12 shows an SEM morphology of dried microcapsules with a core to shell ratio of 1:3 and an average diameter of 25 μm.

The film thickness of asphalt mortar in between aggregate particles in an asphalt mixture is found to be ~ 50 μm [39]. Therefore, in order to avoid being squeezed or pulverised during the asphalt pavement mixing and compaction process, the size of the capsule needs to be less than 50 μm. Su, Qiu also argued that microcapsules of 10 μm and smaller are unsuitable for self-healing, as they are not able to contain sufficient rejuvenators to rejuvenate the aged binder.

The size of the capsule can be controlled by regulating the core to shell ratio (weight ratio between core and shell material). This can be controlled by both

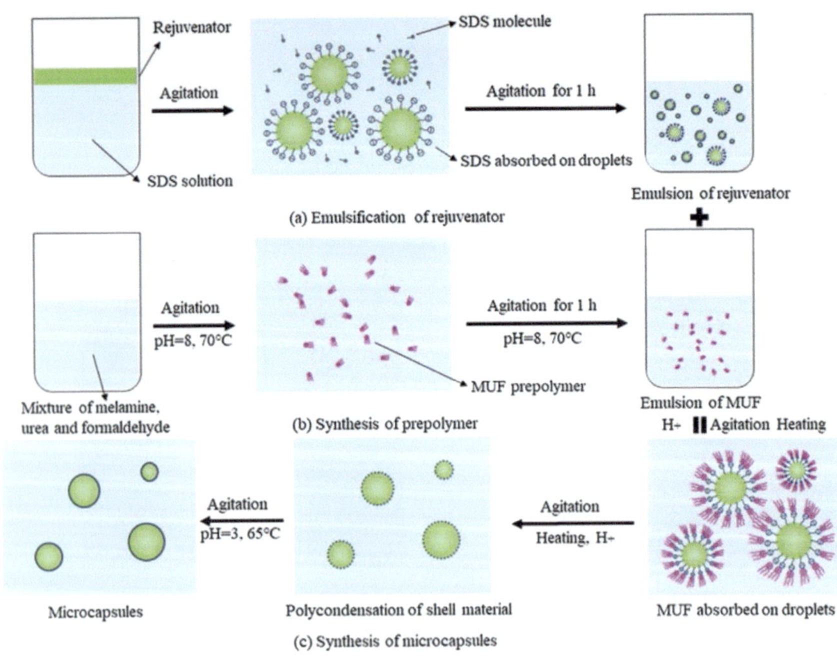

Fig. 11 Sketch mechanism of the fabrication process of the microcapsules containing rejuvenator by in-situ polymerisation (Reproduced from Sun et al. [56])

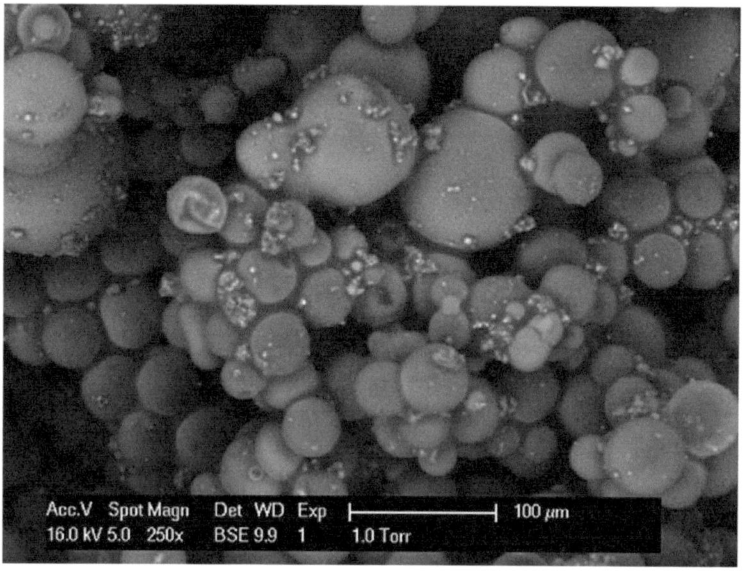

Fig. 12 SEM morphologies of polymer-based microcapsules containing rejuvenator (Reproduced from Su et al. [39])

the prepolymer addition speed and the emulsion stirring rate. Figure 13 shows the morphology of asphalt binder containing varying capsule content (10–30% of total binder volume) and size (10–30 μm). Figure 13 a1 and a2 demonstrate that microcapsules with a mean size of less than 10 μm have a trend of congregation due to the electrostatic attraction of tiny particles. In Fig. 13 b1 and b2, microcapsules of 20 μm have a homogeneous distribution. Figure 13c1 and c2 show larger microcapsules (30 μm size) with asphalt binder dispersing content of 10 and 30%. Figure 13c2 shows that large microcapsules will occupy more space where the asphalt binder dispersion content is 30%. This phenomenon may lead to interface separation between microcapsules and the binder, potentially causing unexpected micro-cracks or other structural defects [39, 57]. This, in turn, could result in poor mechanical performance of the asphalt pavement and, in situ, lead to premature pavement failure. In conclusion, the optimum content of microcapsules in asphalt binder needs to be less than 30% of the asphalt binder volume in an asphalt mix [39].

Sun et al. [58] produced Melamine-Urea–Formaldehyde (MUF) microcapsules with sizes of 40–130 μm to enhance the self-healing properties of asphalt binder and mixture. The original direct tensile stress of asphalt binder and the recovered direct tensile stress after healing were measured using the ductility testing equipment, and the direct tensile stress recovery rate was used to characterize the effect of MUF microcapsules on the self-healing property of asphalt binder. As shown in Fig. 14, MUF microcapsules can effectively enhance the healing efficiency of asphalt binder.

In another research, Sun et al. [59] found that the addition of 3% MUF microcapsules doubled the four-point bending fatigue life of an AC-10 asphalt mixture and

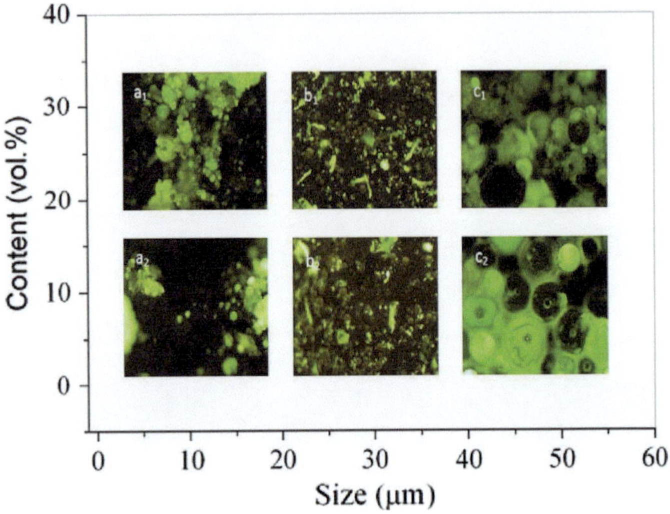

Fig. 13 Fluorescence microscope morphologies of asphalt binder samples with different contents of polymer microcapsules (Reproduced from Su et al. [39])

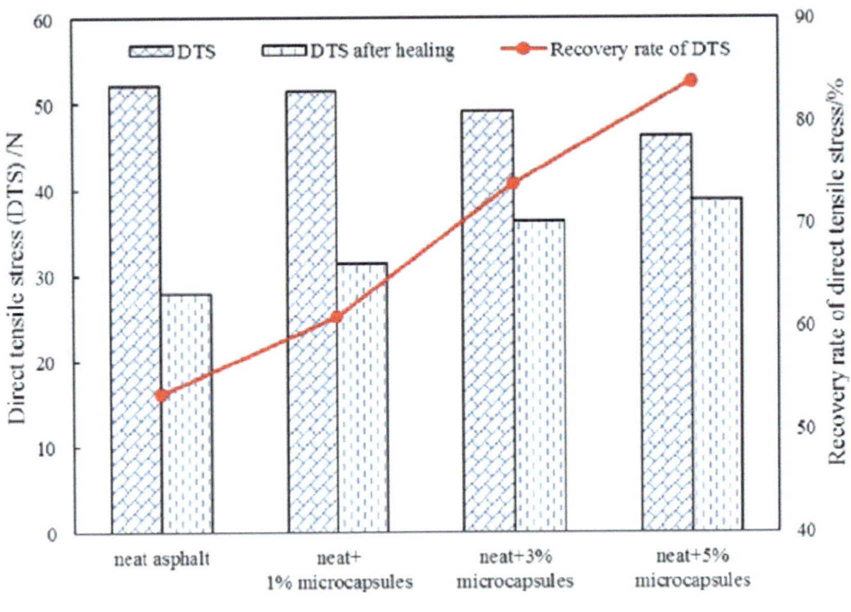

Fig. 14 Tensile strength and recovery ratio of asphalt specimens with different dosages of MUF microcapsules before and after healing (Reproduced from Sun et al. [58])

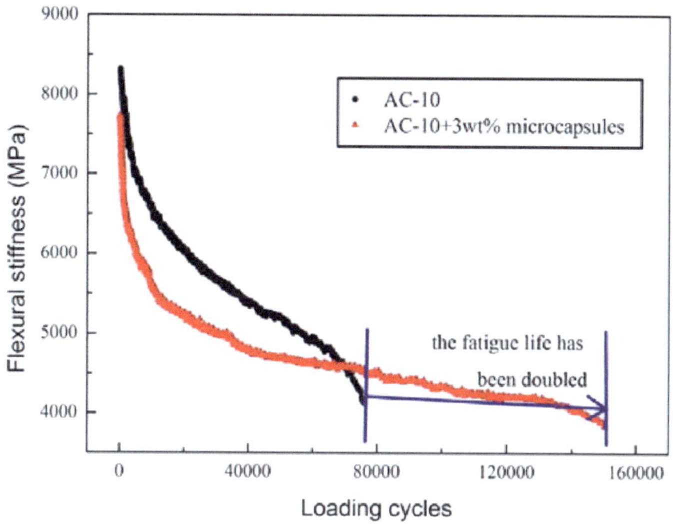

Fig. 15 Fatigue curves of the asphalt mixture specimens with and without MUF microcapsules with a strain of 500 mε (Reproduced from Sun et al. [59])

significantly enhanced the fatigue life recovery of the asphalt mixture after the resting periods as shown in Figs. 15 and 16. Nevertheless, the healing of microcapsules was a one-time thing. Once broken, the microcapsules released the rejuvenator, and the healing effect became very limited when cracks returned.

However, a difficulty with this approach is that large amounts of microcapsules are needed to ensure a sufficient amount of rejuvenator is present in the mixture. The addition of large quantities of microcapsules into the asphalt mixture can reduce the quality of the pavement, which may cause premature pavement failure due to holes in the asphalt binder created by the fracture of microcapsules [60]. Moreover, the chemical compounds used in the production of microcapsules, such as melamine–formaldehyde [61], in large quantities could pose an environmental threat via leaching.

5.4 *Calcium-Alginate Microcapsules*

In the study carried out by Concha et al. [67] they highlighted that for the successful synthesis of the encapsulated rejuvenators in the alginate matrix, their main component, i.e. the oil-in-water (O/W) emulsion, should be physically stable during the encapsulation period. They measured the creaming index of three O/W emulsions, consisting of the measurement of the extent of separation of the two main emulsion components, i.e. (1) the oil: virgin cooking oil (VCO), waste cooking oil (WCO), and virgin engine oil (VEO), and (2) the alginate solution with time, see Fig. 17a.

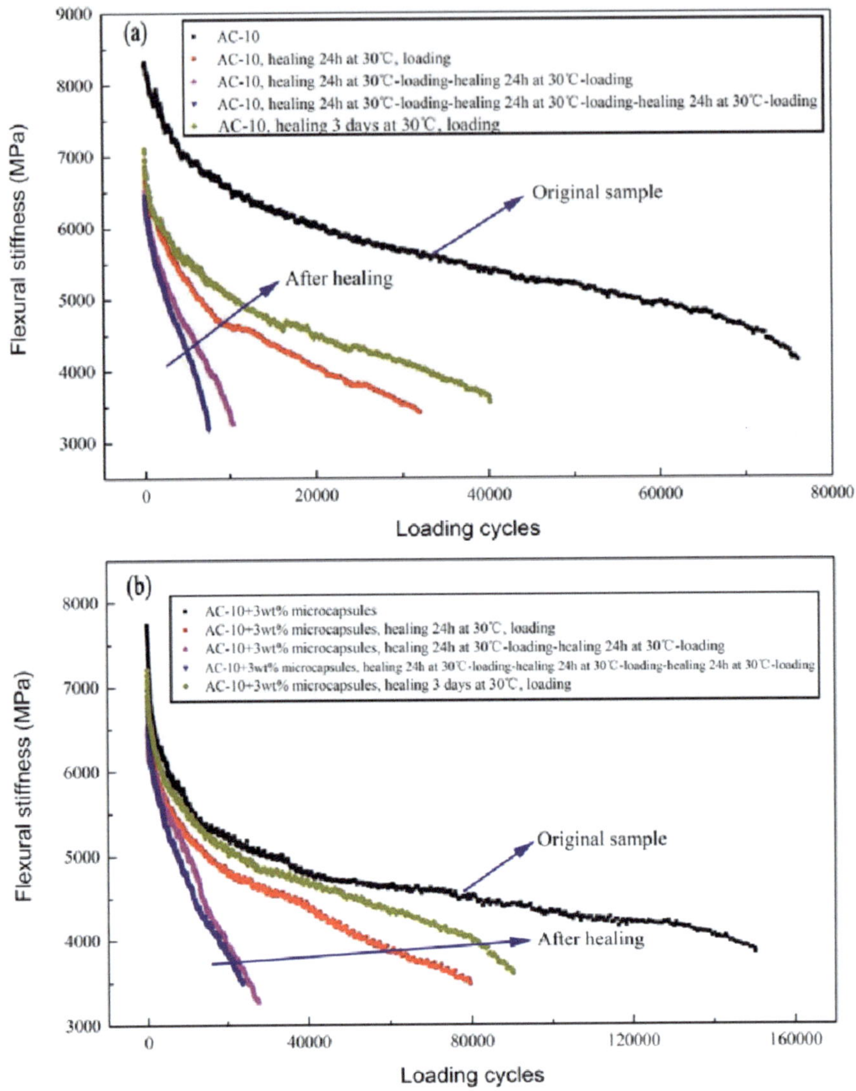

Fig. 16 Fatigue curves of the asphalt mixture at different healing conditions, **a** without microcapsules, **b** with 3 wt% microcapsules (Reproduced from Sun et al. [59])

The main results proved that the higher the creaming index, the higher the extent of separation of the emulsion components, see Fig. 17b. They noted that, overall, the stability of the O/W emulsions was maintained physically stable up to 2 h after synthesis. After this period, the emulsion became physically unstable, causing rapid separation from the alginate solution. This behaviour would be attributed to the fact that the oil droplets, initially dispersed in the alginate solution, start to coalesce,

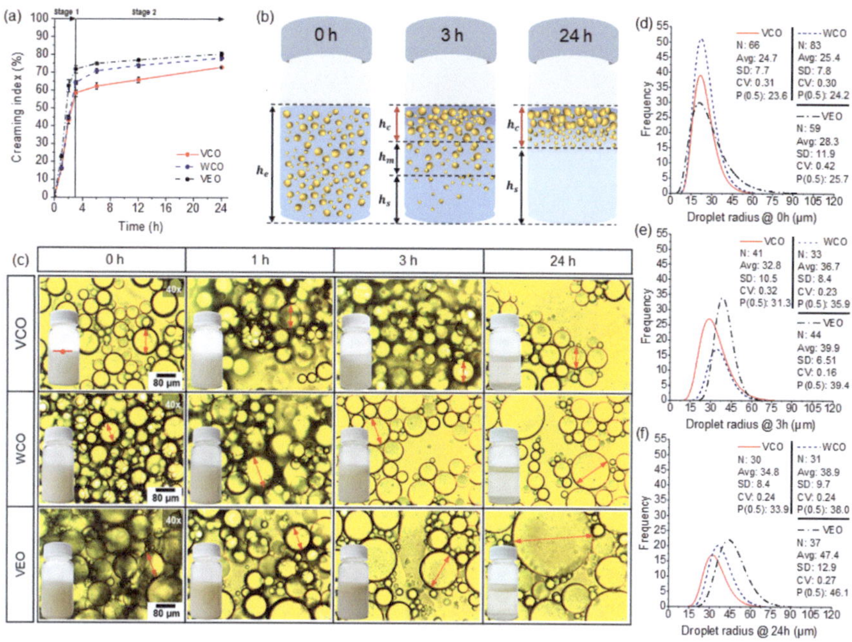

Fig. 17 **a** Average creaming index results for VCO, WCO, and VEO emulsions over time; **b** Representation of gravitational separation occurring in the emulsions by creaming; **c** Microscopy fluorescence images of the droplet size for O/W emulsions over time; and **d–f** Frequency histograms of the droplet size for the O/W emulsions fitted to a log-normal distribution measured at 0, 3, and 24 h (Reproduced from Concha et al. [67])

increasing their size, see Fig. 17c–f. This favors the separation of the oil from the alginate solution. Thus, for encapsulation purposes, they recommended that the encapsulation of alginate-based O/W emulsions should take place for a period no longer than 2 h after synthesization, regardless of the type of oil encapsulated.

Norambuena-Contreras et al. [70] conducted the first study on the synthesis of alginate-based capsules encapsulating pyrolytic oil from end-of-life tyres as a rejuvenator. In this study, authors demonstrated that alginate-based capsules were successfully synthesised via the ionic gelation principle (Fig. 18a). The capsules presented spherical morphology with a representative size of 1.62 mm (Fig. 18b). The Scanning Electron Microscopy (SEM) analysis (Fig. 18d–g) revealed that the pyrolytic oil capsules presented both an external and internal polynuclear structure. The multi-cavity structure of the alginate polymer was able to effectively keep the oil in a way that resulted in encapsulation efficiencies up to 96.54% in the alginate matrix and a representative density of 1.042 g/cm^3 (Fig. 18(h)) which was similar to a virgin asphalt binder (1.044 g/cm^3). Moreover, compressive strength and TGA results on pyrolytic oil capsules and their components confirmed that the encapsulation process of pyrolytic oil as a rejuvenating agent leads to a thermally and mechanically stable material for the extrinsic self-healing of asphalt materials.

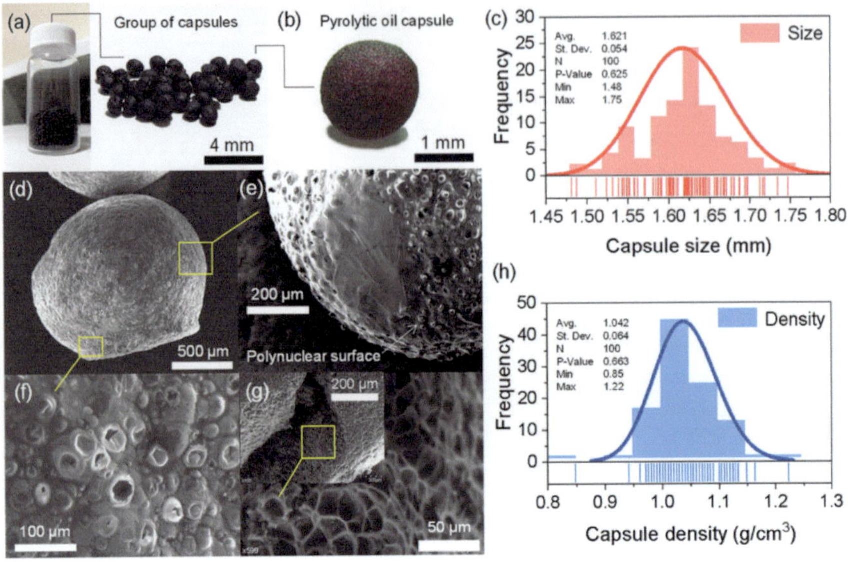

Fig. 18 **a** Pyrolytic oil capsules; **b** individual spherical pyrolytic oil capsule; **c** frequency histogram of the size (diameter) of pyrolytic oil capsules with lognormal fitting; **d** SEM images of pyrolytic oil capsule with a detail of the **e–f** surface and **g** the internal microstructure; and **h** frequency histogram of the relative density of pyrolytic oil capsules with log-normal fitting (Reproduced from Norambuena-Contreras et al. [70])

To address issues of the potential leaching effect of the melamine–formaldehyde capsules, Xu and Tabaković [41] used calcium-alginate as the material for encapsulation of the rejuvenator within asphalt mixture. The authors demonstrated that these capsules have sufficient thermal and mechanical strength to survive the harsh conditions during the asphalt mixing process and can deliver a rejuvenator to the damaged site. Figure 19 shows a microscopic image of the capsule.

Furthermore, Xu and Tabaković [41] reported that the microcapsules do not form a rejuvenator compartment similar to the melamine–formaldehyde capsules but that they form a honeycomb-like structure inside the capsule, where the rejuvenator is encapsulated in several compartments within the capsule. Figure 20 shows the internal structure of the calcium-alginate capsule. This could benefit the self-healing system as it could release the rejuvenator in smaller volumes and, as such, prevents over-saturation of the crack and asphalt mixture. This type of capsule has been called a "polynuclear capsule" by current literature. For more information on this topic, readers are invited to refer to the review on recent advances on self-healing of asphalt materials by the action of encapsulated rejuvenators [37].

Other researchers such as Micaelo et al. [45], Al-Mansoori et al. [46], and Norambuena-Contreras et al. [49] also used calcium-alginate to encapsulate sunflower oil to prepare a variety of self-healing capsules via the dropping funnel method. Micaelo et al. observed large reductions in voids and cracks near the

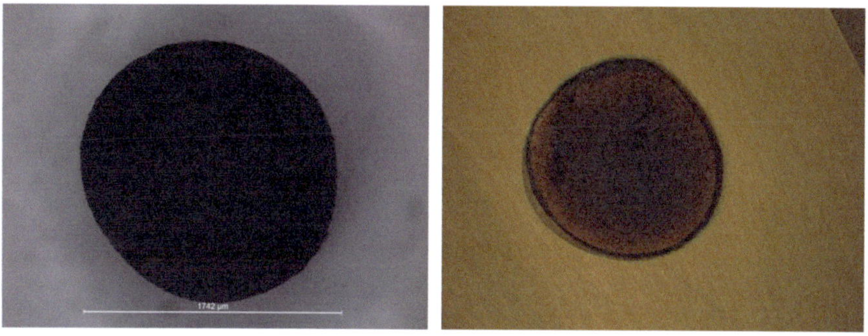

Fig. 19 Microscopic image of calcium-alginate capsule encapsulating rejuvenator; Left: full capsule, Right: cross-sectional image of microcapsule embedded in epoxy (Reproduced from Xu et al. [6])

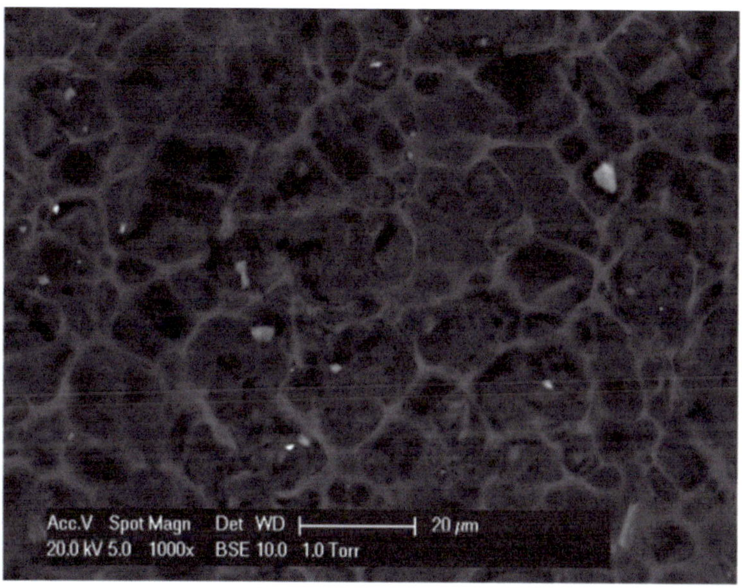

Fig. 20 Environmental scanning microscope image of calcium-alginate capsule internal structure (Reproduced from Xu et al. [41])

capsule's location: oil and/or modified asphalt binder flowed into and filled the cracks to the bottom of the capsules. Al-Mansoori et al. demonstrated that calcium-alginate capsules can significantly enhance the self-healing capacity of asphalt mixture without compromising its mechanical performance, as shown in Figs. 21 and 22, respectively.

Norambuena-Contreras et al. demonstrated that the healing levels of asphalt mixtures with the capsules were higher than the mixtures without the capsules while

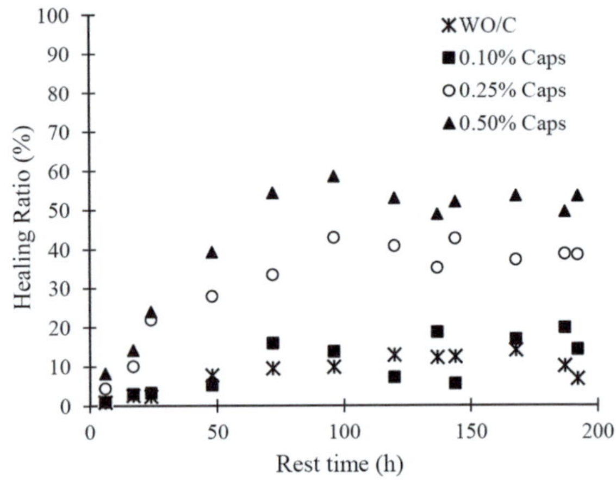

Fig. 21 Healing ratios of asphalt mixtures without and with calcium-alginate capsules (Reproduced from Al-Mansoori et al. [46])

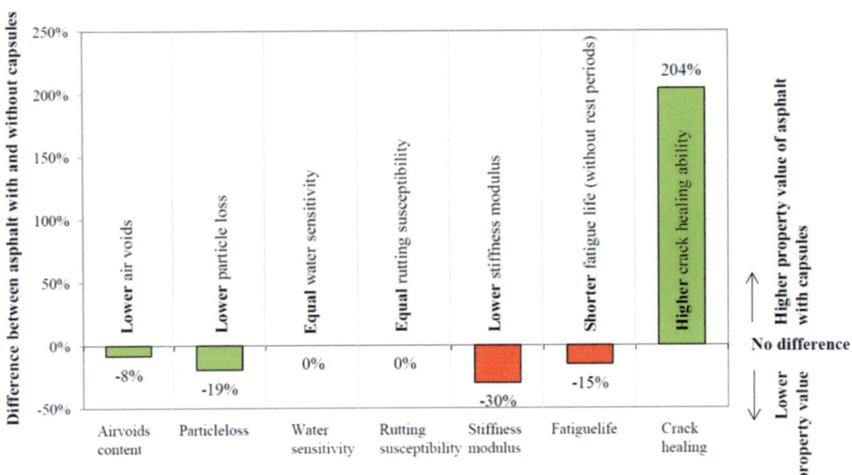

Fig. 22 Performance comparison of asphalt mixtures with and without calcium-alginate capsules (Reproduced from Al-Mansoori et al. [46])

also varying depending on the order of addition and mixtures within the capsules. They recommended adding and mixing the capsules with the asphalt mixture at the end of the mixing process without the need to heat the capsules prior to the mixing process. Zhang et al. [50] found that polynuclear capsules can gradually release sunflower oil from the multi-cavity structure under cyclic compression loading. The sunflower oil released can soften the asphalt mixture, resulting in significant recoveries of the mixture's fracture strength and energy shown in Fig. 23.

Bao et al. [51] found that the incorporation of attapulgite not only enhanced the strength of calcium-alginate capsules but also reduced the premature leakage of the rejuvenator from the capsules, providing asphalt mixture with the ability to release

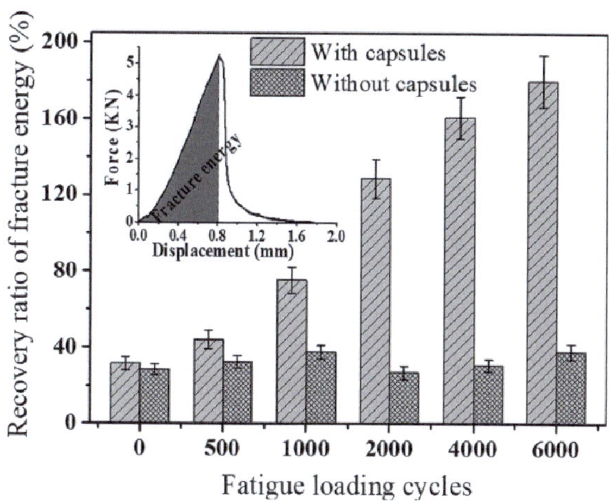

Fig. 23 Fracture energy recovery ratio of asphalt mixture beams after different cycles compression loading at 2.1 MPa (Reproduced from Zhang et al. [50])

the oil gradually to suit the cracking process of asphalt pavement. It was found that the calcium-alginate attapulgite composite capsules within the asphalt mixture could release 55.2% of the oil insides after 64,000 cycles of traffic loading (equivalent to 4 years of service on a medium-traffic grade pavement).

Bao et al. [51] also found that calcium-alginate capsules could gradually release the rejuvenator inside due to shrink and deformation upon cyclic compression loading, with the capsules maintaining their original spherical shape without being fractured after releasing rejuvenating asphalt binder and healing the cracks, as shown in Fig. 24.

Recently, Ruiz-Riancho et al. [52] reported that calcium-alginate capsules may delay reflective cracking, increasing the durability of asphalt up to 21%, but do not

Fig. 24 Capsules taken out of the asphalt mixture specimen after compression loading and healing (Reproduced from Bao et al. [51])

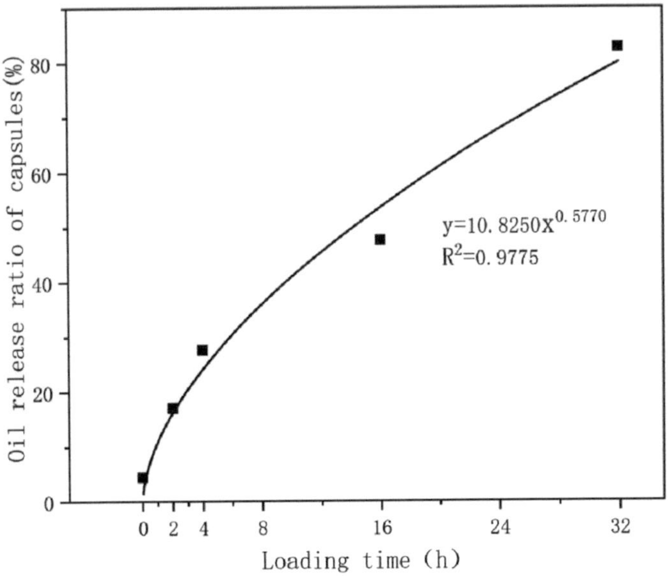

Fig. 25 Oil release ratio of capsules within asphalt slab under wheel tracking at a load pressure 0.7 MPa and a load frequency 60 cycles/min (Reproduced from Rao et al. [62])

show a significant loss in rutting or skid resistance. In this context, Yu et al. [53] reported that calcium-alginate capsules could also improve the self-healing property of aged asphalt mixture. Rao et al. [62] prepared calcium-alginate capsules with accepted properties with industrial-grade sodium alginate, colloid mill, and self-developed titration equipment, which enhanced the capsule preparation efficiency and decreased the raw material cost. They also demonstrated that the oil release ratio of the capsules within the asphalt mixture slab could reach 82.67% after suffering 115,200 cycles of compression loading (equivalent to 7 years of traffic loading on capsules within a medium traffic grade pavement), shown in Fig. 25.

Calcium-alginate capsules have obvious advantages over core–shell structure microcapsules in terms of gradual rejuvenator release patterns and long-term healing potential, making them a promising self-healing technology for asphalt pavement construction and maintenance.

5.5 *Biobased Spore Microcapsules*

In the context of novel biobased encapsulated rejuvenators for asphalt healing, Alpizar-Reyes and coworkers [63] have developed a first study on the use of highly porous spores as an encapsulation media for rejuvenators with asphalt self-healing purposes. They used Sporopollenin spores coming from Lycopodium *clavatum*.

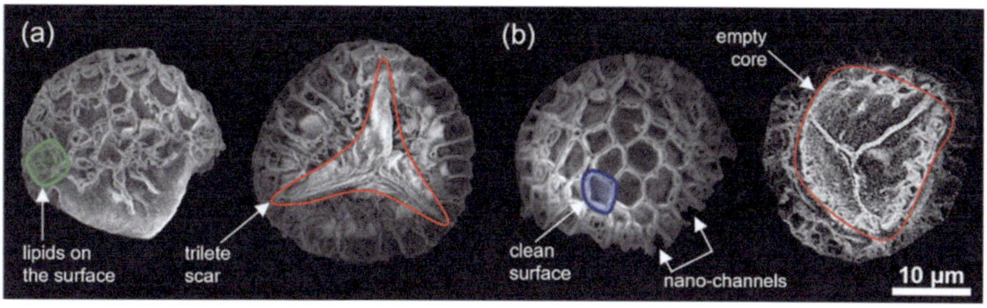

Fig. 26 SEM images of **a** natural spore and **b** treated spore showing empty lycopodium spore (Reproduced from Alpizar-Reyes et al. [63])

These spheres have diameters between 25 and 30 μm, and include a porous network of narrow channels and interconnected pores of ~200 nm, providing abundant space to potentially encapsulate an asphalt rejuvenator loading. Additionally, Sporopollenin is biocompatible and resilient against high temperatures, changes in pH, ionic strength, UV light, oxygen, and mechanical stress, making them suitable for encapsulation. Figure 26 shows representative scanning electron microscopy images of representative of Sporopollenin spores.

Alpizar-Reyes et al. made the first proof of concept of the self-healing ability of the spores by saturating them with sunflower oil as an asphalt binder rejuvenator via vacuum encapsulation. Then, they evaluated the crack-closure process of a PAV-aged asphalt binder sample with a 150 artificial micro-crack width based on previous work published by Norambuena-Contreras et al. [64]. They proved that the spore microcapsules loaded with the sunflower oil could close the micro-crack after 50 min of evaluation (see Fig. 27a, b). The new encapsulation mechanism can promote the redistribution of asphaltenes in a PAV aged asphalt binder, recovering their original agglomeration state and encouraging complete self-healing as seen by the SARA fraction of the restored asphalt binder in Fig. 27c.

5.6 *Compartmented Calcium-Alginate-Fibres*

Tabaković et al. [42] reported a concept where compartmented alginate fibres encapsulating rejuvenator was tested as self-healing technology for asphalt pavements. Figure 28 illustrates a compartmented alginate fibre encapsulating asphalt binder rejuvenator.

Shu et al. [72] innovated the preparation technology and used microfluidic equipment (as shown in Fig. 29) to prepare compartmented calcium-alginate and calcium alginate/silica self-healing fibres encapsulating the rejuvenator. Based on the method, they also produced calcium alginate/graphene oxide with both cracking and microwave response properties [73]. Moreover, they adjusted the flow speed of the rejuvenator and alginate solution to produce calcium alginate fibres with different

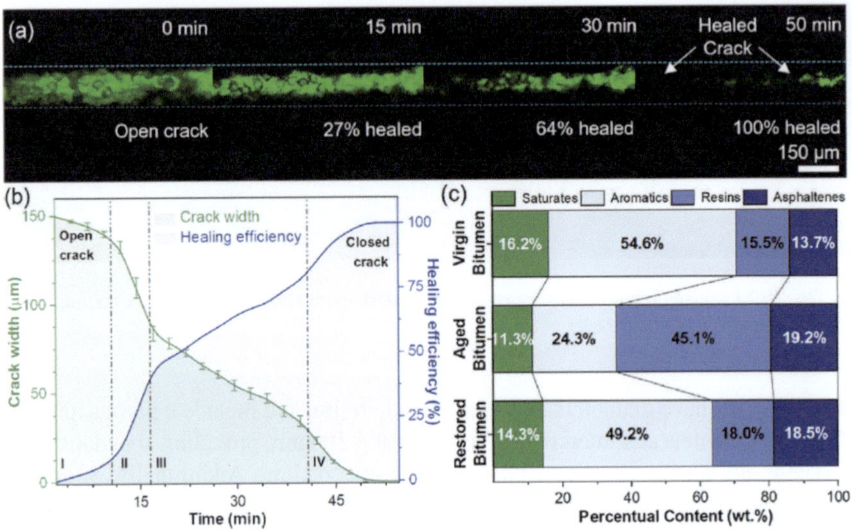

Fig. 27 Asphalt self-healing using biobased spore microcapsules: **a** fluorescence microscope images showing the asphalt crack healing by effect of the oil released from the spores, **b** Curves of crack width and healing efficiency of the closure of the artificial crack in the PAV aged asphalt binder over time; and **c** SARA fractions for the virgin, aged, and rejuvenated asphalt binder samples (Reproduced from Alpizar-Reyes et al. [63])

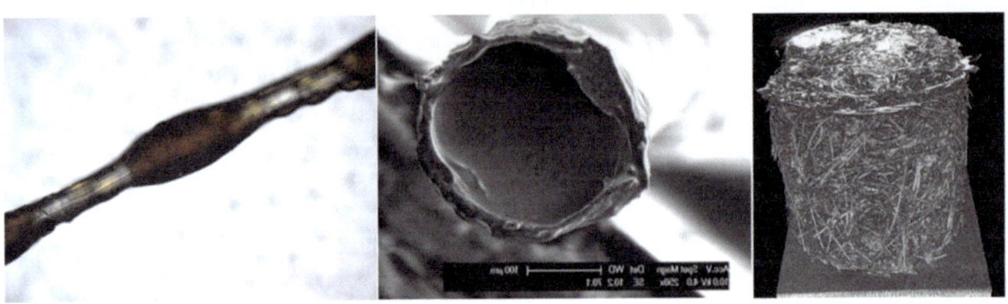

a. b. c.

Fig. 28 Compartmented alginate fibre, **a** compartmented alginate fibre encapsulating rejuvenator, **b** ESEM image of fibre cross-section, and **c** CT scan of alginate fibre distribution throughout asphalt binder test specimen (Reproduced from Tabakovic et al. [42, 71])

inner structures [74]. Following this process, Li et al. [75] fabricated three types of modified calcium-alginate fibre containing soybean oil and optimised the preparation parameters.

Due to the rejuvenator encapsulated in different compartments, fibres can gradually release the inner healing agent sequentially via the microcracks in order to have multiple healing episodes. Shu et al. [74] used the fluorescence microscope to monitor the specific self-healing process of asphalt binder with compartmented calcium-alginate fibres (see Fig. 30) and proposed that the process could be divided

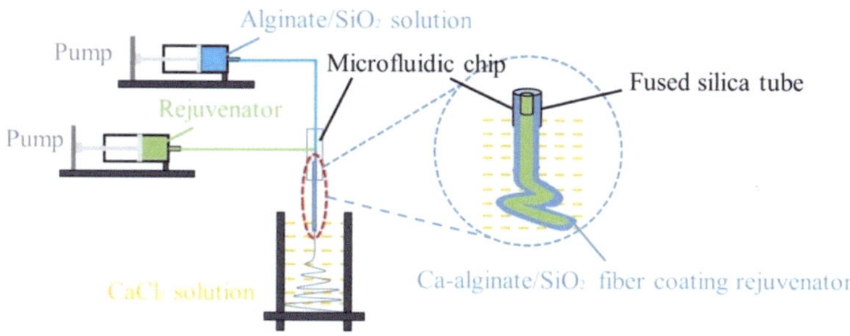

Fig. 29 Schematic of preparation of calcium-alginate fibres by microfluidic device (Reproduced from Shu et al. [72])

into five steps: (1) rupture of inner compartments under the activation of microcracks; (2) rejuvenator released from fibres diffuse and fill in the microcracks due to the capillary flow effect; (3) rapid softening of asphalt binder and sharp decline in asphalt viscosity; (4) the asphalt flows and fills towards the cracked area; (5) closure of microcracks and recovery of property of asphalt. Overall, the rejuvenator release mechanism of hollow and compartmented fibres is a crack-response rupture release mechanism, and the released oil can help the damaged asphalt rapidly complete the healing process.

Tabaković, Post [42] showed that alginate fibres have great potential as a self-healing technique for asphalt pavements as they can be inserted into the asphalt

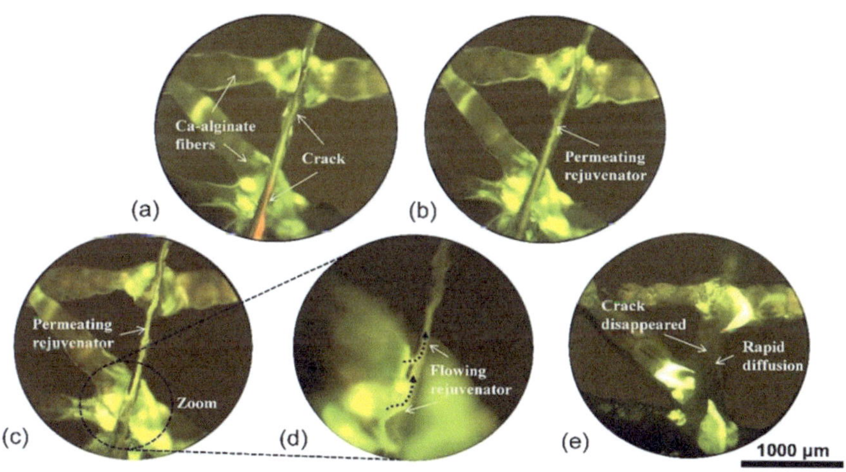

Fig. 30 Self-healing microscopic process of asphalt binder with Ca-alginate fibres characterised by fluorescence microscopy at 15 °C: **a** cracked fibre after 10 s; **b** cracked fibre after 30 s; **c** cracked fibre after 1 min; **d** zoom cracked fibre after 1 min; and **e** cracked fibre after 4 h (Reproduced from Shu et al. [74])

mixtures and increase asphalt mastic mix strength by 36%. Figures 31 and 32 illustrate healing efficiency of an asphalt mortar mixture, with and without fibres. This was explained by the low amount of rejuvenator contained in the mixture; the fibre-rejuvenator alginate ratio was 40:60. Further study showed that increasing the rejuvenator alginate ratio in the fibre mixture to 70:30 improved the healing rate. Tabaković and Braak [54] also showed that amounts of fibre greater than 10% could reduce the healing capacity of the asphalt mixture. However, the study revealed that the fibers have limited healing capacity, with the optimum rejuvenator-to-alginate ratio determined to be 70:30 [71]. While the fibres are effective in healing micro-cracks [42], the system is less efficient in healing large cracks.

Figure 33 shows that the optimum amount of fibre in the mixture was 5% by volume of the constituents. However, further fracture studies demonstrated that even with optimum fibre design and fibre content, the healing of large cracks is hardly improved. Nevertheless, the study showed that alginate-based fibres can successfully heal micro-cracks within an asphalt mixture.

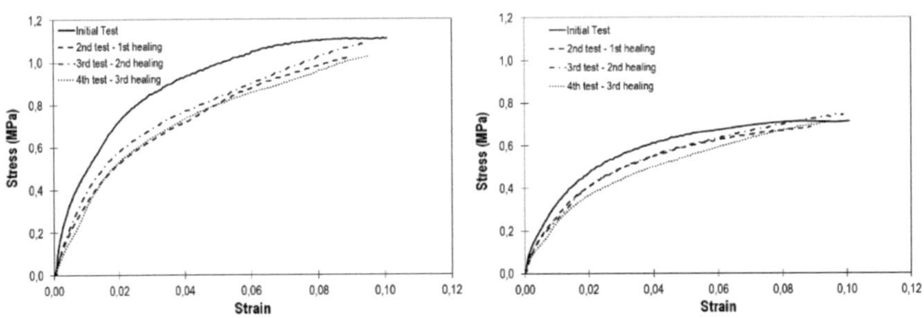

a. b.

Fig. 31 Typical 3PB test load versus deflection plots for asphalt mastic mixture at 20 °C, at varying healing stages: **a** mastic mixture containing fibres, and **b** mastic mixture without fibres (Reproduced from Tabakovic et al. [42])

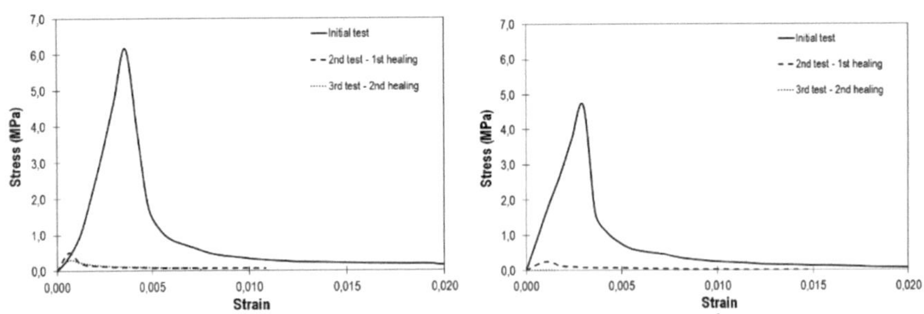

a. b.

Fig. 32 Typical 3PBT load versus deflection plots for asphalt mastic mix at – 5 °C, at varying healing stages: **a** asphalt mixture containing fibres, and **b** asphalt mixture without fibres (Reproduced from Tabakovic et al. [42])

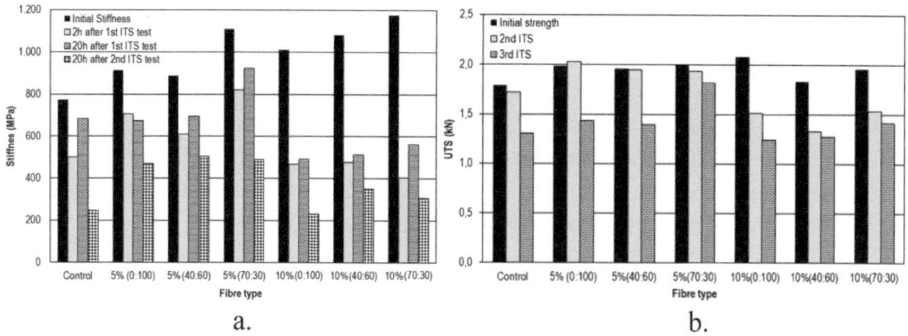

Fig. 33 Healing efficiency of Porous Asphalt (PA) mix containing varying amounts of fibres in the mix **a** stiffness and **b** tensile strength (Reproduced from Tabakovic et al. [71])

Shu et al. [74] conducted tensile stress recovery and modified ductility tests on asphalt binders with hollow and compartmented calcium-alginate fibres. Results indicated that the hollow fibres enhanced their self-healing ability better than compartmented fibres due to more rejuvenators encapsulated in them. Moreover, they also produced compartmented calcium-alginate fibres modified with graphene oxide.

The main test results showed that the healing ratio of a neat asphalt binder sample was 55.3%, the healing index of asphalt binder containing 5% calcium alginate/ graphene oxide fibres reached 74.6%, and with 10 s microwave heating, the healing index increased to 96.1%. It revealed that the self-healing ability of asphalt binder containing compartmented calcium alginate fibres was significantly improved under the synergistic action of rejuvenator and microwave heating [73].

6 Extrinsic Hybrid Asphalt Self-Healing System

Hybrid self-healing systems provide an alternative approach to the challenges associated with induction, microwave heating, and encapsulated rejuvenator technologies. Hybrid self-healing technology combines induction heating and encapsulated rejuvenation. In this process, induction heating is used to repair the asphalt damage (and the cracks), while rejuvenation is used to replenish the aged asphalt binder. Xu et al. [82] demonstrated that a hybrid healing system improves the healing efficiency, allowing for rapid damage repair and a simultaneous rejuvenation of the aged asphalt. However, a challenge with this approach is that the asphalt mix must be adjusted to accommodate both capsules and steel fibres. A solution to this challenge lies in combining the induction and rejuvenation into one product.

Tabaković et al. [83] have developed a novel extrinsic self-healing asphalt technology based on conductive alginate capsules encapsulating asphalt binder rejuvenator (oil). This innovation can reinforce the asphalt pavement by improving its tensile and compressive strength and enhance its autonomous healing properties via induction heating. The concept combines two existing self-healing asphalt technologies: (i) rejuvenator encapsulation and (ii) induction heating to create a self-healing system which provides rapid and effective asphalt pavement repair. Initial results demonstrated that capsules have sufficient thermal and mechanical strength to survive the asphalt mixing process and sufficiently heal the asphalt binder damage [83]. Initial results also show that the capsules can improve the mechanical properties of the asphalt mixture in terms of improving tensile strength and stiffness. With a capsule content of 20% in virgin asphalt binder, the strength of the material increases to 118%. Meanwhile, for asphalt mortar (asphalt binder + fine aggregates – sand), the mixture strength increases by up to 67%. However, the system does not demonstrate a good healing response for asphalt mixture. Although the system is still in progress, it is necessary to determine the optimal amount of capsules in the mix to initiate induction healing while not reducing the mechanical and physical properties of the asphalt mixture.

Wan et al. [84, 85] developed Calcium alginate/nano-Fe_3O_4 composite capsules for controlled rejuvenator oil release using microwave heating. They have shown that Fe_3O_4 alginate capsules encapsulating rejuvenators can successfully be inserted into an asphalt mixture and distributed evenly throughout it. They also showed that the capsules demonstrated superior levels of healing with microwave heating due to the combined effect of thermal induction and rejuvenator healing. The hybrid Fe_3O_4 Calcium alginate capsules encapsulating the rejuvenator have a higher self-healing capacity (8.6%) when compared to the mixture containing pure calcium alginate capsules encapsulating the rejuvenator and 19.3% in comparison to the standard asphalt mixture. The asphalt mixture containing 2% of the capsules achieved up to 87.2% healing after 40 s of healing time. These results show that hybrid asphalt self-healing systems are the most efficient extrinsic healing systems for pavements.

7 Measuring Extrinsic Self-Healing Efficiency

Although research has demonstrated the potential benefit of improved asphalt pavement performance by incorporating extrinsic self-healing technology, it has not yet been universally accepted by the road industry. One of the biggest obstacles to its adoption is the absence of suitable standard testing procedures which could be used to evaluate the healing performance of the asphalt mixture.

Garcia et al. [47] and Su et al. [39] studied a modified Three Point Bend (3PB) test system for the evaluation of an asphalt mixture containing capsules encapsulating a rejuvenator. Tabaković et al. [42] and Norambuena-Contreras and Garcia [29] also used a 3PB test system to evaluate the self-healing effect. The difficulty with this approach is in the asphalt mixture; the mixture is binder-rich and has the potential to heal due to intrinsic healing. Additionally, testing must be performed at low temperatures to avoid material deformation.

Garcia et al. [48] and Barrasa et al. [86] evaluated the performance of asphalt mixtures when subjected to Indirect Tensile Strength (ITS), Indirect Tensile Stiffness

Modulus (ITSM), and Indirect Tensile Fatigue (ITF) testing. However, these studies only reported on the comparative performance between asphalt mixtures with and without healing agents rather than the healing process itself. Tabaković and Braak [44] studied the self-healing performance of Porous Asphalt (PA) mixtures using ITS and ITSM tests. This study demonstrated that the test approach is effective, although ITS testing damages test samples to the extent that they are unsuitable for repeated testing. Furthermore, the ITS test system, the sample thickness, and the dark surface of the test specimens preclude observations of crack propagation and healing unless damage to the test specimen is severe.

More recently, Tabaković and Schuyffel [87] reported a study on the efficiency evaluation of an asphalt mixture healing employing both (i) standard test methods for asphalt pavements such as the ITS and the 4 Point Bending (4PB) Fatigue Test; and (ii) alternative fracture tests such as the Semi-Circular Bending test (SCB). The study employed fracture theory to evaluate the efficiency of the damage repair crack propagation. The study specifically examined the compartmented calcium-alginate fibres encapsulating asphalt binder healing agent (i.e., asphalt binder rejuvenator). The compartmented fibres were used to distribute the rejuvenator (healing agent) throughout the asphalt mix. The research findings demonstrated that the 4PB Fatigue test is the most suitable for evaluating the performance of self-healing asphalt. The form and testing process initiates minor damage (micro-cracks) to the pavement mixture, allowing the test specimen to keep its shape and to be used repeatedly. However, the results also indicated that the ITS and the SCB tests are unsuitable testing methods, as the damage caused to the test specimens precluded repeat testing. 50% of test specimens were not usable after the second test due to the change in their geometrical structure.

Crack healing in asphalt materials using iron-based particles was demonstrated by double torsion tests in model asphalt samples by Jeoffroy and Bouville [88]. With the help of double torsion measurements, they showed that 80% of the original fracture toughness could be recovered by heating the asphalt sample.

Al-Mansoori and Micaelo [46] reported on a novel mechanical method used to measure the crack-healing efficiency of asphalt specimens containing calcium-alginate capsules. This method allows simulating the capsule damage subjected to repeated loading, such as simulating traffic loading. In this work, the self-healing of asphalt mixtures was quantified as the healing level reached from the flexural strength recovery of cracked asphalt mixture beams tested under a three-point bending (3PB) test after a defined healing (rest) time. The proposed crack-healing test considered the following steps: (i) a crack generation in the asphalt samples, (ii) activation of the capsules by compression force, and (iii) rest period and measure of the healing index. The test proved successful in evaluating the healing efficiency of asphalt specimens with and without capsules.

While these experimental approaches are suitable for proof of concept, more robust testing methods are required to convince the road industry to adopt self-healing asphalt material as standard within the road design process. In this sense, Bueno and Arraigada [89] evaluated the healing performance of an induction healing system by employing a large-scale test using the model mobile load simulator MMLS3. The accumulated damage was successfully monitored using the Digital Image Correlation system. The crack initiation detected with this optical technique was taken as a critical limit before irreversible damage would be reached. This study observed that the healing treatment carried out by induction heating allowed specimens to recover their initial performance. It was also proven that the number of loading cycles until total failure was longer when the healing process was carried out before the damage started. In conclusion, this experimental method was found useful to confirm that the strength was recovered after a healing process. Hence, an increase in the service life of the asphalt pavement might be obtained.

8 Towards New Generations of Extrinsic Self-Healing Asphalt Pavements

The key aim of self-healing asphalt pavements is to develop asphalt pavement material capable of healing without external intervention. Therefore, the last goal for highway designers is to create an asphalt pavement material that will mimic nature itself. To achieve this, the self-healing processes embedded within the asphalt pavement system should be able to self-assess themselves. This would enable the material to monitor the health of its structure and materials at any time and location and respond in a manner that fulfils its functional requirements based on the information gained [90]. Authors such as Tabaković and Schlangen [91] and, more recently, Gonzalez-Torre and Norambuena-Contreras [37] stated that to develop a new generation of self-healing asphalt pavements, based on findings of currently available extrinsic self-healing technologies, presented above, three specific working areas are identified that need particular effort:

(1) **Development/design of damage sensing and repair triggering elements**. To be incorporated within the asphalt pavement system, the elements will be able to trigger the self-healing process (i.e., signalling transport and activation of the healing mechanism). This means that the sensory function must be enhanced and extended with an active learning functionality, able to differentiate and detect damage, interpret the obtained information and trigger/stimulate the healing action on demand. These sensor elements should ideally be structural components of the pavement system and should not deteriorate the general functionality of the pavement system. The development of the sensory mechanism within the pavement system will allow for healing-on-demand action; such an action could be triggered by a fall in current/resistivity in the pavement system or by a concentration of stress, which will initiate the repair action while activating an initiator (healing agent or heating). The healing process would be activated instantly without delay, ensuring fast damage repair (providing that external conditions are adequate, i.e., healing time and temperature). As such, the materials would remain damage-free.

(2) **Development of multiple self-healing processes**. Only a limited number of self-healing mechanisms for asphalt pavements have been developed, as illustrated in Sects. 5.3, 5.4, 5.5 and 5.6. To meet the functional requirements of pavements, new self-healing mechanisms must be developed and employed to cover a broader range of performance demands, such as more realistic healing/rest time. All four self-healing mechanisms presented in the previous sections require at least 4 h of rest period for a mixture to achieve recovery. On busy highways, a rest period of four hours is not expected to occur naturally. The self-healing times for asphalt pavement need to be improved. Another essential part of the self-healing mechanism in asphalt pavements is multi-healing action. For instance, once a core–shell structure capsule is broken and the rejuvenator is released, no further healing actions can be expected, and the asphalt is no longer self-healing. This will ultimately lead to asphalt pavement failure.

(3) **Development of self-healing assessment techniques**. Carry out autonomous health assessment of the asphalt pavement system (i.e., a self-monitoring system) and, in case of damage repair, quantify the success of the self-healing process. To date, there has been only a limited understanding of the quantification of the success of self-healing, mostly by measurement of mechanical performance, such as material strength. This requires on-site pavement evaluation or obtaining test samples from site and conducting laboratory material evaluation. This involves traffic control measures, which could result in traffic delays and high financial and environmental costs. A mechanism for the autonomous assessment of asphalt pavement system health and the assessment of the self-healing process needs to be a focus point of future research in this field.

Suppose these developments in the self-healing processes are met. In that case, it will be possible to create a truly smart asphalt pavement system that senses its internal state and external environment and responds appropriately to this information. The primary advantage of moving towards smart/self-healing technology is the potential cost benefit of condition-based maintenance strategies and the prospective lifespan that may be achieved for asphalt pavement materials through *in-situ* health management. The potential benefits of self-healing asphalt technology in material performance, environmental, and social benefits will undoubtedly stimulate interest in the wider use of self-healing technology in road design and construction. However, for self-healing technology to become accepted as the industry standard, its superiority in the construction and maintenance of asphalt pavements must be demonstrated, from a functional perspective, as well as from economic and environmental approaches.

9 Summary

Incorporating extrinsic self-healing technology into the asphalt pavement design may present a solution to some of the difficulties facing asphalt. Existing technologies have demonstrated the potential to repair the distressed asphalt pavements. They offer great opportunities for increased durability and reliability, reduced maintenance, and overall costs of asphalt pavements. This includes diminished material resources since the usual over-design of materials is no longer required. The repair of asphalt pavement is addressed in situ by its internal self-healing system at the very position of the first appearance of damage, eliminating the need for a classical in-situ maintenance process.

Despite the progress made in the development of self-healing asphalt technology, further work is required to focus on:

i. damage sensing and repair triggering element,
ii. the development of multiple self-healing processes, and
iii. the development of reliable self-healing assessment techniques.

References

1. Qiu, J., et al. (2009). Investigation of self-healing capability of bituminous binders. In *Road materials and pavement design; Special issue on asphalt materials (ICAM 2009—China)* (Vol. 10, no. 1, pp. 81–94).
2. Khattak, M. J., Khattab, A., Zhang, P., Rizvi, H. R., & Pesacreta, T. (2013). Microstructure and fracture morphology of carbon nano-fiber modified asphalt and hot mix asphalt mixtures. *Materials and Structures, 46*, 2045–2057.
3. Khattak, M. J., Khattab, A., & Rizvi, H. R. (2013). Characterization of carbon nano-fiber modified hot mix asphalt mixtures. *Construction and Building Materials, 40*, 738–745.
4. Santagata, E., Baglieri, O., Tsantilis, L., & Chiappinelli, G. (2015). Fatigue properties of bituminous binders reinforced with carbon nanotubes. *International Journal of Pavement Engineering, 16*(1), 80–90.
5. Brown, J. M., Anderson, D. P., Justice, R. S., Lafdi, K., Belfor, M., Strong, K. L., et al. (2005). Hierarchical morphology of carbon single-walled nanotubes during sonication in an aliphatic diamine. *Polymer, 46*, 10854–65.
6. Fenner, J. S., & Daniel, I. M. (2014). Hybrid nanoreinforced carbon/epoxy composites for enhanced damage tolerance and fatigue life. *Composites A, 65*, 47–56.
7. Montazeri, A., & Chitsazedeh, M. (2014). Effect of sonication parameters on the mechanical properties of multi-walled carbon nanotube/epoxy composites. *Materials and Design, 56*, 500–508.
8. Wang, Z. J., Kwon, D. J., Gu, G. Y., Kim, H. S., Kim, D. S., Lee, C. S., et al. (2013). Mechanical and interfacial evaluation of CNT/polypropylene composites and monitoring of damage using electrical resistance measurements. *Composite Science Technology, 81*, 69–75.
9. Fang, C., et al. (2013). Nanomaterials applied in asphalt modification: A review. *Journal of Material Science and Technology, 29*(7), 589–594.
10. Liu, G., van de Ven, M., Wu, S., Yu, J., & Molenaar, A. (2011). Influence of organomontmorillonites on fatigue properties of bitumen and mortar. *International Journal of Fatigue, 33*, 1574–1582.

11. Wu, S., Wang, J., & Jiesheng, L. (2010). Preparation and fatigue property of nanoclay modified asphalt binder. In *International conference on mechanic automation and control engineering (MACE).*
12. Tabatabaee, N., & Shafiee, M. H. (2012). Effect of organoclay modified binders on fatigue performance. In *7th RILEM international conference on cracking in pavements.* RILEM Bookseries.
13. Liu, G. *Characterization and identification of bituminous materials modified with montmorillonite nanoclay.* The Netherlands. ISBN 978-90-8570-779-0.
14. Kim, Y. R., Little, D. N., & Benson, F. C. (1990). Chemical and mechanical evaluation on healing mechanism of asphalt concrete. *Journal of Association of Asphalt Paving Technologists, 759*, 240–275.
15. Santagata, E., Baglieri, O., Tsantilis, L., Dalmazzo, D., & Chiappinelli, G. (2016). Fatigue and healing properties of bituminous mastics reinforced with nano-sized additives. *Mechanics of Time-Dependant Materials, 20*, 367–387.
16. Manh, H. T., & Viet, A. P. (2013). Influence of fiber polymer reinforced asphalt concrete pavement in high temperature environment. In *2nd international interdisciplinary conference.* Slovak Republic.
17. McNally, T. (2011). Ch. 2: Polymer modified bitumen, PmB. In *Polymer modified bitumen properties and characteristics.* Woodhead Publishing.
18. Abtahi, S. M., Sheikhzadeh, M., & Hejazi, S. M. (2010). Fiber-reinforced asphalt-concrete—A review. *Construction and Building Materials, 24*, 871–877.
19. Liu, Q. (2012). *Induction healing of porous asphalt concrete.* Faculty of Civil Engineering and Geosciences. TU Delft.
20. García, A. (2012). Self-healing of open cracks in asphalt mastic. *Fuel, 93*, 264–272.
21. Liu, Q., et al. (2012). Evaluation of the induction healing effect of porous asphalt concrete through four point bending fatigue test. *Construction and Building Materials, 29*, 403–409.
22. Liu, Q., et al. (2010). Induction heating of electrically conductive porous asphalt concrete. *Construction and Building Materials, 24*, 1207–1213.
23. Dai, Q., Wang, Z., & Mohd Hasan, M. R. (2013). Investigation of induction healing effects on electrically conductive asphalt mastic and asphalt concrete beams through fracture-healing tests. *Construction and Building Materials, 49*(Supplement C), 729–737.
24. García, Á., et al. (2012). Optimisation of composition and mixing process of a self-healing porous asphalt. *Construction and Building Materials, 30*, 59–65.
25. García, A., et al. (2009). Electrical conductivity of asphalt mortar containing conductive fibers and fillers. *Construction and Building Materials, 21*(10), 3175–3181.
26. García, A., et al. (2013). Induction healing of dense asphalt concrete. *Construction and Building Materials, 49*, 1–7.
27. Schlangen, E. (2013). Other materials, applications and future developments. In M. Rooij, et al. (Eds.), *Self-healing phenomena in cement-based materials* (pp. 241–256). RILEM Series: State-of-the-Art Reports.
28. Schlangen, E., et al. (2011). Highway A58: The first engineered self-healing asphalt road. In *3rd international conference on self-healing materials.* Bath.
29. Norambuena-Contreras, J., & Garcia, A. (2016). Self-healing of asphalt mixture by microwave and induction heating. *Materials & Design, 106*, 404–414.
30. Gallego, J., et al. (2013). Heating asphalt mixtures with microwaves to promote self-healing. *Construction and Building Materials, 42*, 1–4.
31. Collin, R. E. (2000). *Foundations for microwave engineering* (2nd ed.) Willey-IEE Press.
32. Norambuena-Contreras, J., & Gonzalez-Torre, I. (2017). Influence of the microwave heating time on the self-healing properties of asphalt mixtures. *MDPI Applied Sciences, 7*(10), 15.
33. Sun, Y. H., et al. (2014). Microwave heating of steel slag asphalt mixture. *Key Engineering Materials, 599*, 5.
34. Benedetto, A., & Calvi, A. (2013). A pilot study on microwave heating for production and recycling of road pavement materials. *Construction and Building Materials, 44*, 9.

35. González, A., et al. (2018). Effect of RAP and fibers addition on asphalt mixtures with self-healing properties gained by microwave radiation heating. *Construction and Building Materials, 159*(Supplement C), 164–174.
36. Trigos, L., Gallego, J., Escavy, J. I., & Picado-Santos, L. (2021). Dielectric properties versus microwave heating susceptibility of aggregates for self-healing asphalt mixtures. *Construction and Building Materials, 293*, 123475.
37. Gonzalez-Torre, I., & Norambuena-Contreras, J. (2020). Recent advances on self-healing of bituminous materials by the action of encapsulated rejuvenators. *Construction and Building Materials, 258*, 119568.
38. García, Á., Schlangen, E., & van de Ven, M. (2010). Two ways of closing cracks on asphalt concrete pavements: Microcapsules and Induction Heating. *Key Enginering Materials, 417–418*, 573–576.
39. Su, J. F., et al. (2014). *Experimental investigation of self healing behaviour of bitumen/microcapsule composites by modified beam on elastic foundation method*. Springer Publication, RILEM.
40. Su, J. F., & Schlangen, E. (2012). Synthesis and physicochemical properties of high compact microcapsules containing rejuvenator applied in asphalt. *Chemical Engineering Journal, 198–199*, 289–300.
41. Xu, S., et al. (2016). *Alginate capsules encapsulating rejuvenator as self-healing system for porous asphalt concrete*. HelCON.
42. Tabaković, A., et al. (2016). The reinforcement and healing of asphalt mastic mixtures by rejuvenator encapsulation in alginate compartmented fibres. *Smart Materials and Structures, 25*(8).
43. Tabakovic, A., et al. (2017). An evaluation of the efficiency of compartmented alginate fibres encapsulating a rejuvenator as an asphalt pavement healing system. *MPDI Applied Sciences, 7*, 16.
44. Tabaković, A., et al. (2017). The compartmented alginate fibres optimisation for bitumen rejuvenator encapsulation. *Journal of Traffic and Transportation Engineering (English Edition)*.
45. Micaelo, R., Al-Mansoori, T., & Garcia, A. (2016). Study of the mechanical properties and self-healing ability of asphalt mixture containing calcium-alginate capsules. *Construction and Building Materials, 123*(Supplement C), 734–744.
46. Al-Mansoori, T., et al. (2017). Microcapsules for self-healing of asphalt mixture without compromising mechanical performance. *Construction and Building Materials, 155*(Supplement C), 1091–1100.
47. Garcia, A., Austin, C. J., & Jelfs, J. (2016). Mechanical properties of asphalt mixture containing sunflower oil capsules. *Journal of Cleaner Production, 118*, 9.
48. Garcia, A., Jelfs, J., & Austin, C. J. (2015). Internal asphalt mixture rejuvenation using capsules. *Construction and Building Materials, 101*, 8.
49. Norambuena-Contreras, J., et al. (2018). Effect of mixing and ageing on the mechanical and self-healing properties of asphalt mixtures containing polymeric capsules. *Construction and Building Materials, 175*, 254–266.
50. Zhang, L., et al. (2019). Synthesis and characterisation of multi-cavity Ca-alginate capsules used for self-healing in asphalt mixtures. *Construction and Building Materials, 211*, 298–307.
51. Bao, S., et al. (2020). Synthesis and characterisation of calcium alginate-attapulgite composite capsules for long term asphalt self-healing. *Construction and Building Materials, 265*, 120779.
52. Ruiz-Riancho, N., et al. (2021). Properties of Ca-alginate capsules to maximise asphalt self-healing properties. *Construction and Building Materials, 284*, 122728.
53. Yu, X., et al. (2022). Effect of ageing on self-healing properties of asphalt concrete containing calcium alginate/attapulgite composite capsules. *Materials, 15*(4), 1414.
54. Tabaković, A., et al. (2015). The use of compartmented Sodium-Alginate fibres as healing agent delivery system for asphalt pavements. In *European Materials Research Society Fall Meeting, Warsaw, Poland*.
55. Su, J. F., Qiu, J., & Schlangen, E. (2013). Stability investigation of self-healing microcapsules containing rejuvenator for bitumen. *Polymer Degradation and Stability, 98*(6), 1205–1212.

56. Sun, D., et al. (2018). Optimisation of synthesis technology to improve the design of asphalt self-healing microcapsules. *Construction and Building Materials, 175*, 88–103.
57. Su, J. F., et al. (2011). Interface stability behaviors of methanol– melamine–formaldehyde shell microPCMs/epoxy matrix composites. *Polymer Composites, 32*, 810–820.
58. Sun, D., et al. (2017). Enhanced self-healing process of sustainable asphalt materials containing microcapsules. *ACS Sustainable Chemistry & Engineering, 5*(11), 9881–9893.
59. Sun, D., et al. (2018). Fatigue behavior of microcapsule-induced self-healing asphalt concrete. *Journal of Cleaner Production, 188*, 466–476.
60. Li, R., Zhou, T., & Pei, J. (2015). Design, preparation and properties of microcapsules containing rejuvenator for asphalt. *Construction and Building Materials, 99*, 143–149.
61. Anderson, F. A. (1995). Final report on the safety assessment of melamine/formaldehyde resin. *Journal of the American College of Toxicology, 14*(5), 373–385.
62. Rao, W., et al. (2021). Efficient preparation and characterisation of calcium alginate-attapulgite composite capsules for asphalt self-healing. *Construction and Building Materials, 299*, 123931.
63. Alpizar-Reyes, E., Concha, J. L., Martín-Martínez, F. J., & Norambuena-Contreras, J. (2022). Biobased spore microcapsules for asphalt self-healing. ACS Appl. Mater. Interf. **14**, 31296–31311.
64. Norambuena-Contreras, J., Arteaga-Perez, L. E., Guadarrama-Lezama, A. Y., Briones, R., Vivanco, J. F., & Gonzalez-Torre, I. (2020). Microencapsulated bio-based rejuvenators for the self-healing of bituminous materials. *Materials, 13*(6), 1446.
65. Xu, S., Liu, X., Tabaković, A., Lin, P., Zhang, Y., Nahar, S., Lommerts, B. J., & Schlangen, E. (2021). The role of rejuvenators in embedded damage healing for asphalt pavement. *Materials and Design, 202.*
66. Su, J. F., et al. (2015). Investigation the possibility of a new approach of using microcapsules containing waste cooking oil; in-situ rejuvenation. *Construction and Building Materials, 74*, 83–92.
67. Concha, J. L., Arteaga-Pérez, L. E., Alpizar-Reyes, E., Segura, C., Gonzalez-Torre, I., Kanellopoulos, A., & Norambuena-Contreras, J. (2022). Effect of rejuvenating oil type on the synthesis and properties of alginate-based polynuclear capsules for asphalt self-healing. *Road Materials and Pavement Design, 24*, 1669–1694.
68. Norambuena-Contreras, J., Concha, J. L., Arteaga-Pérez, L. E., & Gonzalez-Torre, I. (2022). Synthesis and characterisation of alginate-based capsules containing waste cooking oil for asphalt self-healing. *Applied Sciences, 12*(5), 2739.
69. Erdogan Yamaç, Ö., Yilmaz, M., Yalçın, E., Kök, V., Norambuena-Contreras, J., & Garcia, A. (2021). Self-healing of asphalt mastic using capsules containing waste oils. *Construction and Building Materials, 270*, 121417.
70. Norambuena-Contreras, J., Arteaga-Pérez, L. E., Concha, J. L., & Gonzalez-Torre, I. (2021). Pyrolytic oil from waste tyres as a promising encapsulated rejuvenator for the extrinsic self-healing of bituminous materials. *Road Materials and Pavement Design, 22*(S1), S117–S133.
71. Tabaković, A., et al. (2017). The compartmented alginate fibres optimisation for bitumen rejuvenator encapsulation. *Journal of Traffic and Transportation Engineering, 4.*
72. Shu, B., et al. (2018). Synthesis and characterisation of compartmented Ca-alginate/silica self-healing fibers containing bituminous rejuvenator. *Construction and Building Materials, 190*, 623–631.
73. Shu, B., et al. (2019). Synthesis and properties of microwave and crack responsive fibers encapsulating rejuvenator for bitumen self-healing. *Materials Research Express, 6*(8), 085306.
74. Shu, B., et al. (2019). Microfluidic synthesis of polymeric fibers containing rejuvenating agent for asphalt self-healing. *Construction and Building Materials, 219*, 176–183.
75. Li, Y., Hao, P., & Li, N. (2022). Preparation and properties of a novel rejuvenator-loaded fiber for asphalt pavement. *Construction and Building Materials, 324*, 126687.
76. Gerardu, J. J. A., & Hendriks, C. F. (1985). *Recycling of road pavement materials in the Netherlands.* Road Engineering Division of Rijkswaterstaat.
77. Kandhal, P.S., et al. (1995). *Performance of recycled hot mix asphalt mixtures.* National Center for Asphalt Technology.

78. Wu, S., et al. (2007). Investigation of temperature characteristics of recycled hot mix asphalt mixtures. *Conservation and Recycling, 51*, 610–620.
79. Silva, H. M. R. D., Oliveira, J. R. M., & Jesus, C. M. G. (2012). Are totally recycled hot mix asphalts a sustainable alternative for road paving? *Resources, Conservation and Recycling, 60*, 38–48.
80. Brownridge, J. (2010). The role of an asphalt rejuvenator in pavement preservation: use and need for asphalt rejuvenation. In *1st International conference on pavement preservation*. Newport.
81. Tabaković, A. G., A.; McNally, C., & Gilchrist, M. D. (2010). The influence of recycled asphalt pavement on the fatigue performance of asphalt concrete base courses. *ASCE Journal of Materials in Civil Engineering, 22*(6), 643–650.
82. Xu, S., et al. (2020). A novel self-healing system: Towards a sustainable porous asphalt. *Journal of Cleaner Production, 259*, 120815.
83. Tabakovic, A., Mohan, J., & Karac, A. (2021). Conductive compartmented capsules encapsulating a bitumen rejuvenator. *MPDI Processes, 9*(8).
84. Wan, P., et al. (2021). A novel microwave induced oil release pattern of calcium alginate/nano-Fe_3O_4 composite capsules for asphalt self-healing. *Journal of Cleaner Production, 297*, 126721.
85. Wan, P., et al. (2021). Self-healing properties of asphalt concrete containing responsive calcium alginate/nano-Fe_3O_4 composite capsules via microwave irradiation. *Construction and Building Materials, 310*, 125258.
86. Barrasa, R. C., et al. (2014). Addressing durability of asphalt concrete by self-healing mechanism. *Procedia - Social and Behavioral Sciences, 162*, 188–197.
87. Tabaković, A., et al. (2017). An evaluation of the efficiency of compartmented alginate fibres encapsulating a rejuvenator as an asphalt pavement healing system. *MDPI Applied Sciences, 7*(7), 16.
88. Jeoffroy, E., et al. (2018). *Iron-based particles for the magnetically-triggered crack healing of bituminous materials. Construction and Building Materials—Currectly under the review.*
89. Bueno, M., Arraigada, M., & Partl, M. N. (2016). Damage detection and artificial healing of asphalt concrete after trafficking with a load simulator. *Mechanics of Time-Dependent Materials, 20*(3), 265–279.
90. Fisher, H. (2010). Self repairing materials—Dream or reality. *Natural Science, 2*(8), 873–901.
91. Tabaković, A., & Schlangen, E. (2016). Self-healing technology for asphalt pavements. In M. D. Hager, S. van der Zwaag, & U. S. Schubert (Eds.), *Self-healing materials* (pp. 285–306). Springer International Publishing.

Experimental Characterization of Self-Healing of Asphalt Materials

Orazio Baglieri, Fabrizio Miglietta, Greet Leegwater, Amir Tabaković, Lucia Tsantilis, Roberto M. Aurilio, and Hassan Baaj

Abstract Several efforts have been made by the research community to identify proper laboratory testing methods and develop various analytical interpretations to estimate the healing performance of asphalt materials at the binder, mastic, mortar, and mixture scales. Depending on the considered scale and the testing approach, the evaluation of healing can capture two distinct mechanisms: adhesive healing and cohesive healing. While adhesive healing considers the phenomenon at the binder-aggregate interface, cohesive healing refers to the rebonding action of the binder medium itself. The assessment of crack healing can be derived from a variety of testing modes used to induce damage in the material and allow for the self-healing phenomenon to manifest itself when the material is left at rest. In such a context, damage can be based on both fatigue or fracture modes, requiring different experimental protocols and theoretical approaches. When focusing on the rest phase, the testing methods can be generally classified as storage, intermittent and hybrid healing tests, depending on how rest period(s) are introduced in the overall testing protocol. Following these considerations, several healing parameters can be introduced to measure the healing performance of the asphalt materials.

O. Baglieri (✉) · F. Miglietta · L. Tsantilis
Department of Environment, Land and Infrastructure Engineering, Politecnico di Torino, 24, c.so Duca degli Abruzzi, 10129 Turin, Italy
e-mail: orazio.baglieri@polito.it

G. Leegwater
Netherlands Organisation for Applied Scientific Research (TNO), NEXT Delft, Delft, The Netherlands

A. Tabaković
Revolution Road Ltd, Wicklow, Ireland

School of Civil Engineering, University College Dublin, Belfield, Dublin, Ireland

R. M. Aurilio · H. Baaj
Department of Civil and Environmental Engineering, University of Waterloo, 200 University Ave W, N2L 3G1, Waterloo, ON, Canada

© The Author(s), under exclusive license to Springer Nature Switzerland AG 2025

H. Baaj and O. Baglieri (eds.), *Crack-Healing of Asphalt Pavement Materials*, RILEM State-of-the-Art Reports 45, https://doi.org/10.1007/978-3-031-99443-2_4

1 Adhesive Versus Cohesive Approaches

Two main types of crack propagation and healing mechanisms are recognized to occur within asphalt materials, referred to as adhesive and cohesive processes [11, 30, 61, 74]. The former considers the crack and healing phenomena that occur at the binder-aggregate interface due to the rebonding action of the binder phase to the mineral aggregate. The latter refers to the crack and healing phenomena that occur at the binder medium's microcrack faces only.

The adhesive failure at the binder-aggregate interface, in combination with cohesive failure in the binder medium, leads to severe decay of both the durability and mechanical performance of asphalt mixtures. Loss of adhesion typically involves thin film coatings of asphalt binder all around the aggregate surface, whereas loss of cohesion occurs in the case of thick film coatings [13]. Thus, the adhesive properties of the binder-aggregate bond and the cohesive properties of the binder exert an important influence on the service life of asphalt pavements [74].

To characterize adhesive healing, it is necessary to study the binder-aggregate system [13, 23]. Adhesion occurring at the binder-aggregate interface must be investigated to understand cracking and healing in asphalt mixtures comprehensively. When focusing on the scale of mastics, mortars and mixtures, the adhesive properties of the binder-aggregate system cannot be discriminated against those related to the cohesion of the asphalt matrix. In this context, researchers have implemented the binder bonding strength (BBS) test to grasp the bonding healing at the binder-aggregate interface. Williams and co-workers [74] explained the cohesive and adhesive bonding interactions of binder-aggregate systems through the Lifshitz-van der Waals interactions of electron shells of neighboring molecules and acid–base interactions between constituent molecules. Bhasin and Little [13] proposed a methodology to estimate the surface free energy components of aggregates using the microcalorimeter. The thermodynamic concepts have been similarly applied by Nejad and co-workers [42] to study the moisture sensitivity of asphalt mixtures in the case of hydrate lime-treated aggregate. In their study, Huang et al. [27] investigated the influence of pulling rate and binder film thickness on the bonding strength of binders mixed with different additives. They found that the pull-off tensile strength (POTS) of the modified binders was improved, and the self-healing ability of blends with crumb rubber and ethylene bis-stearamide (EBS) wax was improved in comparison with that of the base binder. Binder blends with increasing SBS modification levels tended to have higher POTS values but decreased healing rate. Similarly, Lv and co-workers [38] underlined that the SBS modification mostly led to lower adhesive healing percentages than neat asphalt binders. Moreover, they found that the mastic samples made up of limestone filler performed better than binder samples in the investigated healing conditions. Wang et al. [72] used a granite substrate in the presence of water and found overall positive effects on adhesive healing.

Cohesive healing is generally investigated by isolating the asphalt binder phase from the lithic one to exclude any confounding interaction between aggregate and asphalt binder [57, 58, 71]. Such an approach has the advantage of assessing more fundamental properties in the analysis of the overall damage behavior of the asphalt matrix. First detailed by Wool and O'Conner, it was theorized that thermoplastic polymers could heal through the diffusion of molecular chains into contacted crack surfaces [75]. Full healing occurs when the molecular chains fully integrate and randomize into the undamaged polymer matrix, this theory was later adapted to cohesive healing for asphalt materials. Research studies have demonstrated that cohesive healing in asphalt binders results from several processes occurring at crack interfaces, such as wetting, interfacial cohesion, interdiffusion, and randomization of molecules at the wetted interfaces [54]. This allows the crack to be repaired, thus returning the binder matrix to its original or "undamaged" state. Although, from a theoretical point of view, cohesive healing can only be assessed by considering the pure asphalt phase, it is widely recognized that the healing response of the mastic, mortar and mixture differs from that of the associated pure asphalt binder because of the discontinuity introduced by the presence of the aggregate. The cohesive property of the binder is altered by the interaction existing between the binder and the filler/sand fractions [40, 50]. A mixed mode of cohesive and adhesive healing should be considered in such cases. Nevertheless, some researchers also consider the binding part in the mixture (i.e., asphalt binder and filler) as a whole. Williams and co-workers [74] stated that the cohesive fracture can be promoted by high mastic volumes within the asphalt mixture, whereas a lower quantity of asphalt binder can encourage adhesive fracture. Thus, the proper selection of the scale of investigation is of primary importance when studying adhesive and cohesive healing processes. Cohesive healing can be studied on the binder scale only, whereas a mixed approach, comprising both adhesion and cohesion, is involved for the other scales of observation, such as mastic, mortar, and mixture. Alternatively, when investigating only adhesive healing, the interface of the binder-aggregate system should be evaluated.

2 Fatigue-Based Versus Fracture Mechanics Approaches

To assess healing, cracks need to be generated within the asphalt material to subsequently analyze the crack-closure process. Such cracks can originate from both fatigue-based damage and fractures.

Fatigue cracking is the consequence of repeated tensile stresses or strains occurring due to the cyclic passage of traffic wheels during the pavement service life. It occurs at first as microcracks that eventually coalesce to form macrocracks throughout the asphalt layer. Accelerated laboratory testing methods associated with the formation of microcracks are generally referred to as fatigue tests. Fatigue tests are carried out by subjecting the sample to a stress-controlled or strain-controlled load. In stress-controlled tests, the stress input is kept constant, and the strain increases, whereas strain-controlled tests apply constant strain to the sample and the stress is reduced [53]. For the healing assessment, the fatigue-based approach is generally modified by including unloading phases that facilitate the healing phenomena. Fatigue cracks typically initiate as microcracks and grow over time as dictated by the crack propagation process, including the coalescence of the microcracks [22]. When microcracks form macrocracks, fracture-based concepts can be introduced. The healing phenomenon can be studied through fatigue analysis if the microstructural changes are bounded within the domain of microcracks. When the healing phenomenon is considered after the propagation of macrocracks, fracture-based approaches are adopted [34]. Fracture-based tests deal with the complexity of the stress field around the crack tip. The purpose of such tests is to determine the potential course of action of a crack: to propagate or to arrest. This is carried out by considering the fracture energy, which represents the energy required to separate the material into two pieces. Fracture energy is a material-dependent property, measured as the ratio of the work of the fracture to the crack area, where the work of the fracture represents the area under the load–displacement curve [51]. In this context, the assessment of healing is conducted by visibly fracturing the material and by placing the broken pieces back in intimate contact. However, the normal forces applied to bring the two fractured surfaces in contact can cause irregularities to the fractured faces. This can lead to discontinuous contact caused by consequent permanent deformations inevitably occurring at the crack interface [64].

In both approaches, the resistance of the asphalt mixture against cracking depends on several factors, such as the asphalt binder type and content, the volumetric characteristic of the mixture, the aggregate gradation, lithic type and size, as well as the presence of modifiers. Moreover, the adhesive/cohesive forces and the internal friction also dictate the cracking resistance and the healing performance [51]. The two approaches can be applied at different scales, such as asphalt binder, mastic, mortar and mixture, but the basic assumptions for damage and healing are different.

In interpreting the results obtained from fatigue-based and fracture-based test methods, two main research approaches can be adopted: the phenomenological and mechanistic approaches. The phenomenological approach is quite simple, and it relates the fatigue life of the asphalt mixture to the initial pavement response in terms of stress or strain [35]. However, it does not consider the damage evolution process throughout the test, thus being strongly dependent on the material property, test type, and loading conditions [10, 32]. Mechanistic approaches are inherently more complex than the phenomenological approach because they are conceived to be potentially applicable to a broader range of loading and environmental conditions. Such approaches are based on fracture mechanics or continuum damage mechanics [30, 32]. Based on the type of approach adopted for the evaluation of healing, either fatigue-based or fracture-based approaches, the concepts of fracture mechanics or continuum damage mechanics can be implemented. As mentioned, the basic assumption in fatigue cracking is that microcracks are uniformly spread within the material. For such a situation, it is more appropriate to adopt the continuum damage mechanics concepts, which are based on stress–strain relationships. Continuum damage theories can be adopted to model the response of composite materials at different length scales. Those theories allow the characterization of material behavior using the net effect of microstructural changes on observable properties rather than microscale observations [66]. In doing so, they ignore the heterogeneity of the asphalt mixture by hypothesizing that all the damage within the material can be averaged and represented by phenomenologically defined internal state variables [31]. When such assumptions cannot hold due to the presence of a visible crack, it is preferable to use fracture mechanics concepts. The fundamental principles of fracture mechanics underline that cracking initiates in the vicinity of the crack tip in cases where the energy stored nearby exceeds the material crack resistance [59].

A way to bridge the gap between fracture mechanics and continuum damage mechanics is represented by the cohesive zone models used in crack initiation and propagation and, therefore, potentially adopted in the case of crack closure and healing [44]. Such a model implies a more physically sound approach that allows stress singularities ahead of the crack tip to be removed, thus better predicting its advancement during the damaging process [31]. However, the identification of the material constants associated with this approach is still very challenging.

3 Testing Configurations

Many test set-ups have been used to characterize the healing of asphalt binder, mastic, mortar, and mixtures by adopting fatigue-based and fracture-based approaches.

In the context of asphalt binders, cohesive healing is generally investigated through a variety of rheological measurements consisting of fatigue-based testing that all have in common the use of the dynamic shear rheometer (DSR), as presented in Fig. 1a [6, 17, 41]. Such tests are typically conducted on small disk-shaped specimens using a parallel plate configuration, where one DSR plate is fixed, and the other applies a shear force. Dynamic shear tests have also been implemented on two-piece samples to assess the material fracture-related properties. Such an approach is thought to simulate the crack healing process in which the crack is physically reproduced by the two halves of a sample that are brought in intimate contact to trigger the healing phenomena [14]. The direct tension test and the ductility test (Fig. 1b, d) have also been adopted on samples that are preliminarily cut into two equal parts and then placed in contact [48]. This allows the study of fracture-dependent healing behavior in the presence of visible cracks. The cohesive healing/failure of asphalt binders has also been assessed by means of the local fracture test, presented in Fig. 1c. Such a test is performed on an asphalt binder lens placed between two convex protuberances that simulate two aggregates [26].

Laboratory testing methods to investigate the healing behavior associated with the mastic phase have been conducted using dynamic shear tests (Fig. 2e), direct tension tests (Fig. 2g) and three-point bending tests (Fig. 2h). Such testing configurations evaluate the overall healing performances of asphalt mastic samples without separating healing phenomena related to adhesion or cohesion. Similarly, to asphalt binders, dynamic shear tests consist of rheological measurements conducted by imposing small amplitude loading on the sample [56]. Concerning fracture-based approaches, Al-Mansoori and co-workers [3] studied fracture healing on mastic beams that were broken in two pieces using three-point bending tests and then left to heal. In the same context, Qiu et al. [49] implemented the direct tension test to impart

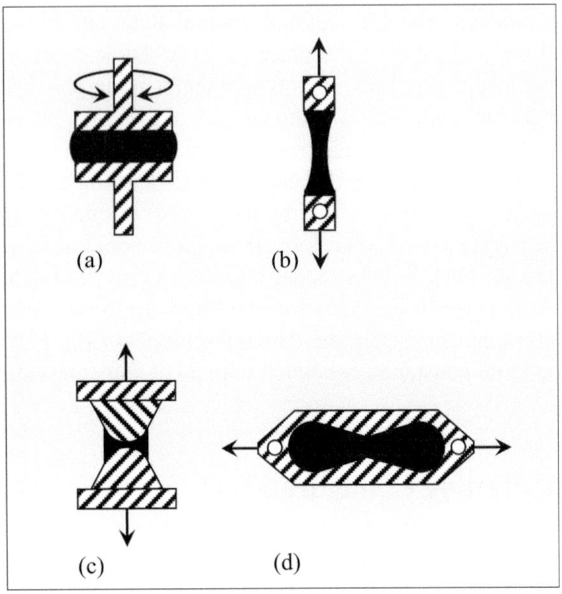

Fig. 1 Schematic representation of testing protocols on the binder scale: **a** dynamic shear test; **b** direct tension test; **c** local fracture test; **d** ductility test

multicracks to the mastic phase before healing. Another recent category of test used for the healing assessment is the binder bonding strength test, schematically presented in Fig. 2f. This configuration can be adopted for both asphalt binder and mastic. The former allows the assessment of the adhesive healing at the binder/aggregate interface [38, 72], while the latter refers to a mixed mode of healing comprising both adhesion and cohesion [27].

When referring to asphalt mortars, laboratory testing has much less tradition in the research community. Despite this, some healing studies are reported in the literature [68, 77]. Similar to the testing configurations adopted in the case of asphalt materials, some of the tests are related to dynamic shear tests and binder bonding strength tests, schematically represented in Fig. 3i, j, respectively. The former allows for the characterization of fatigue-related healing properties, whereas the latter is used in the context of fracture-based approaches. Other testing configurations involve the adoption of the semi-circular bending test and the three-point bending test, displayed in Fig. 3k, l, respectively. Apostolidis et al. [4] implemented the three-point bending test on beam specimens preliminarily broken in two pieces to evaluate the fracture-related healing performances. In the same context of fracture-based approaches, Pan et al. [43] used a notched semi-circular sample to be used for the semi-circular bending test.

When considering the scale of asphalt mixtures, several fatigue-based testing methods predominantly involve the use of cylindrical samples (Uniaxial Tension–Compression (also called pull–push test occasionally) and indirect tensile fatigue tests, pictured in Fig. 4m, n, and prismatic beam samples (three-point bending test and four-point bending test, represented in Fig. 4q, r). Ashouri et al. [5] adopted the

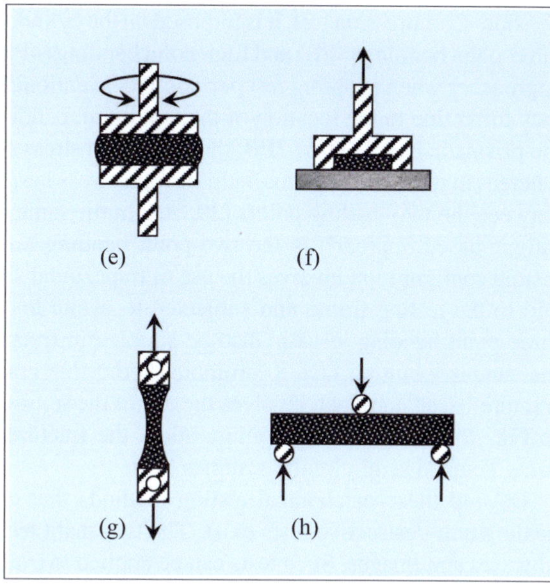

Fig. 2 Schematic representation of testing protocol on the mastic scale and at the binder/mastic-aggregate interface: **e** dynamic shear test; **f** binder bonding strength test; **g** direct tension test; **h** three-point bending test

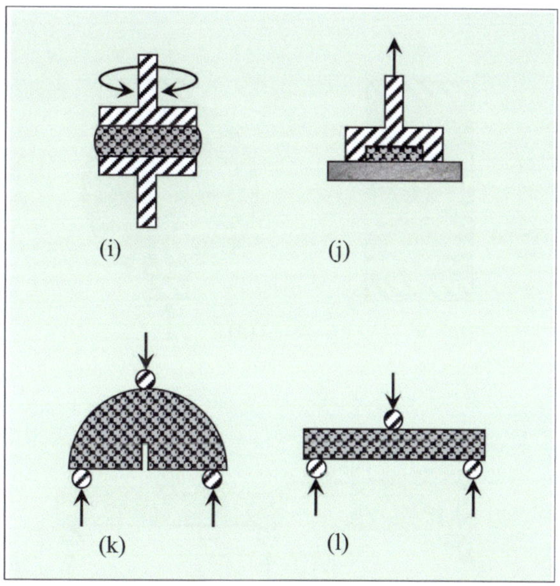

Fig. 3 Schematic representation of testing protocol on the mortar scale: **i** dynamic shear test; **j** binder bonding strength test; **k** semi-circular bending test; **l** three-point bending test

zero-mean stress-controlled testing configuration (Uniaxial Tension–Compression test) in which the haversine loading consisted of a 0.1-s loading pulse with rest periods ranging from 10 to 270 s. Similar loading conditions are adopted for the indirect tensile fatigue test, although the load is applied diametrically, while in the

Tension–Compression test, it is imposed on the cylindrical faces of the specimen. The three-point bending (3PB) and four-point bending (4PB) tests refer to fatigue-healing approaches when adopting rest periods to conventional fatigue procedures. However, they differ due to the location of the maximum bending moment and stresses along the prismatic beam. In the 3PB, the maximum stress occurs under the loading anvil, whereas in the 4PB, the maximum flexural stress is spread all over the beam section between the two loading points [29, 70]. In the same context of beam tests, another fatigue-based approach is the two-point bending test, displayed in Fig. 4p. Such testing configuration involves the use of trapezoidal samples that are clamped at one end to the testing frame and subjected to cyclic loading at its free end [69]. The three-point bending test can also be adopted in fracture-based studies by notching the sample's bottom face to promote predictable crack propagation [21]. Another fracture-based approach involves the use of the semi-circular bending test displayed in Fig. 4o. In this testing configuration, the fracture-healing test is performed by using notched semi-circular samples [29].

Beyond these mechanical testing methods that cause damage to the material, healing non-destructive tests exist. The two main techniques are related to the use of waves and images. Such tests can be applied to evaluate structural changes and to

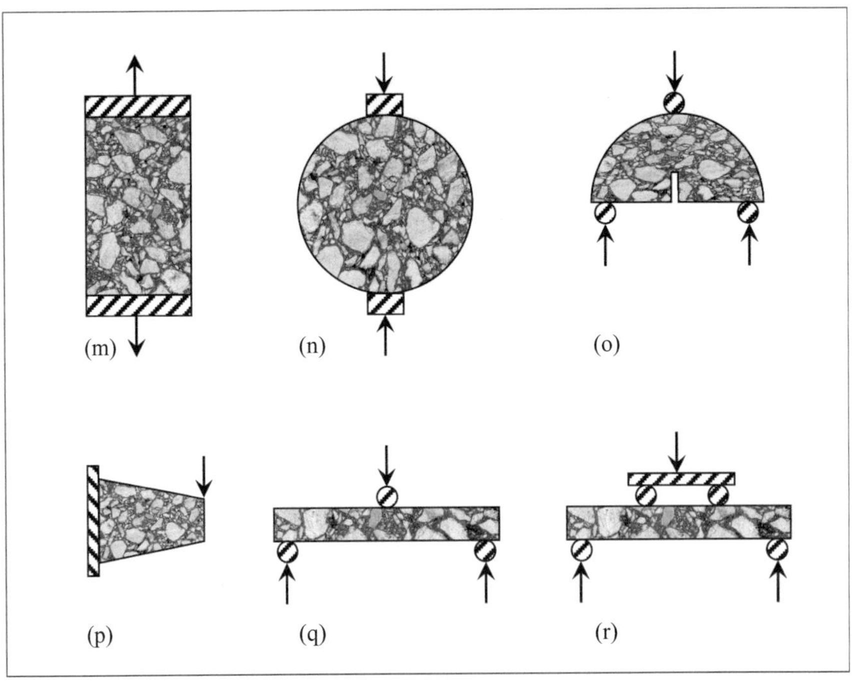

Fig. 4 Schematic representation of testing protocol on the mixture scale: **m** uniaxial Tension–Compression test; **n** indirect tensile test; **o** semi-circular bending test; **p** two-point bending test; **q** three-point bending test; **r** four-point bending test

predict the fatigue life and crack healing of asphalt materials [8]. Studies conducted by [7] and [78] utilized non-invasive ultrasonic propagation measurements to evaluate the fatigue and restoration properties of both unmodified and polymer-modified asphalt mixes using a uniaxial Tension–Compression fatigue-healing test. Ultrasonic transducers were placed axially at both ends of the cylindrical specimens, and a high-frequency (150 kHz) pulse was transmitted through the specimen at regular intervals, the time signals were processed using both time- and frequency-domain methods to evaluate changes in material integrity during loading and rest periods. Using a similar technique, Zhao [78] used the ultrasonic pulse propagation technique coupled with an "interrupted" ideal cracking test (CT) to evaluate the crack-healing ability of asphalt mixes loaded diametrically.

4 Loading Pattern and Rest Periods

Generally, the investigation of healing requires rest period(s) to be introduced during the test. Two different categories of test methods can be distinguished based on how rest period(s) are introduced. One category loads a sample until damaged and then put to rest for a certain period of time. This type of healing test will be referred to as a storage self-healing test. The second category introduces a rest period after each load application in a fatigue test. This type of healing test is referred to as an intermittent self-healing test.

It should also be noted that self-healing can also occur without the introduction of rest periods. In a cyclic test, there is a harmonic load signal. Consequently, the loading varies during the test, and there are periods with decreasing strain or even opposite strain. As a result, cracks that have formed during the largest load peak can be closed later in the loading cycle, which can result in a restoration of strength [36, 45].

The first step of a storage healing test is to introduce (micro)cracks in a specimen. After damage is introduced, the next step is to allow the specimen to rest under controlled conditions: at a certain temperature, for a specific time, and under a certain normal force. After this, the performance of the specimen can be tested and compared to the undamaged performance. It is important to note that, at least in this context, "damage" is intended as a deterioration of the volume of mechanical properties of materials, as it can be evaluated using a continuum mechanics approach. The relation between damage and the formation of microcracks within the material is an interpretation of the observed variation of material properties .

The magnitude and extension of the damage induced to the material can be the result of any kind of cracking introduced to the sample; micro-cracking induced by dynamic fatigue loading [12], a partially cracked sample [26], a fully cracked sample [28, 57] or even by using two separately produced samples [14, 33].

In many cases, a fatigue load is chosen to introduce damage simulating the conditions under which the material would experience wear, before being allowed to heal. An example of the response of a storage healing test is presented in Fig. 5. The black

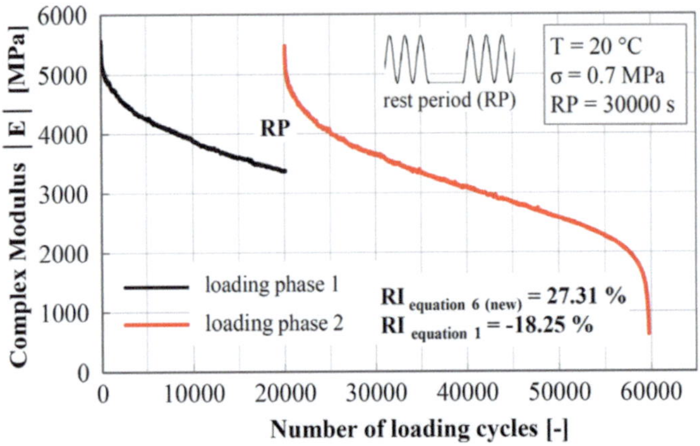

Fig. 5 Stiffness measured in a uniaxial tension–compression fatigue test with a rest period of 30.000 s [28]

line in the figure represents the stiffness reduction due to repeated loading. After a certain stiffness reduction, a rest period (RP) of 30,000s (~ 8 h) is introduced. The red line shows the response of the specimen after the test is started again. The initial stiffness in the second loading phase, just after the rest period, is almost the same as the initial stiffness in the first loading phase. The red line has a slightly higher inclination compared to the black line, which could imply that damage development proceeds faster in an already damaged specimen.

Such a loading condition has been adopted in fatigue-based dynamic shear tests (Figs. 1a and 2e), for asphalt binders and mastics, Tension–Compression test, indirect tensile test, three-point bending test, and four-point bending test (Fig. 4m, n, q, r) to assess the healing performance of asphalt mixtures.

With intermittent healing tests, the purpose is to create a loading pattern which matches the loading in practice more closely. This is carried out by introducing a rest period after each loading period. By definition, this type of protocol is part of a fatigue test. The rest period is usually a multiple of the loading period, such as 10 times the loading period. Such a loading condition was adopted in dynamic shear test, in the case of asphalt binders and mastics, and for the asphalt mixture uniaxial Tension–Compression test, indirect tensile test, and four-point bending test (Figs. 1a, 2e, 4m, n, r).

A disadvantage of this type of test is that it is very time-consuming. Normal fatigue tests are already time-consuming, as one test takes up multiple hours. By also adding a rest period, which multiplies the test time by at least a factor of 10, a single test may take several days. While these types of protocols are considered relevant for research purposes, they are not suitable for inclusion in pavement design protocols. It is also a challenge to apply a representative load signal in between the rest periods due to the load control technology in place in a testing machine for cyclic loading, as these machines often require a number of loading cycles to realize the prescribed

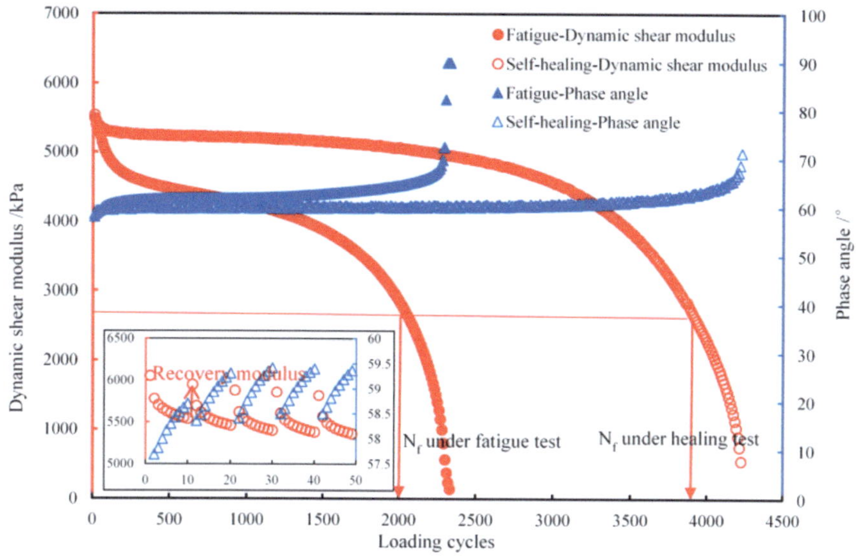

Fig. 6 Stiffness measured in dynamic shear fatigue test with intermittent rest periods of 2 s [60]

load level [67]. With these types of tests it is very important to double-check if the applied load signal is similar to the prescribed load signal as it may affect test results and lead to wrong conclusions.

An example of the response of an intermittent healing test is presented in Fig. 6. The two red lines in the figure represent the stiffness reduction as a result of pulse cycles and pulse-rest cycles. In the case of healing, the applied rest period is 2 s. The comparison among the red lines shows the gain in the material resistance due to healing.

The literature also presents hybrid healing tests, which are a combination of both healing test types, storage and intermittent healing tests. These protocols include multiple loading cycles followed by multiple rest cycles. In some cases, a hybrid loading regime is required as test equipment cannot realize one loading signal with several rest periods with sufficient accuracy. This was the case for tests performed by Van den Bergh [68]. To address this, he selected a protocol with 30 loading cycles followed by 90 rest cycles, giving him a loading-to-healing ratio of 3. However, a hybrid protocol can also aim to introduce not 1 large but 10 small rest periods of 2 min during a fatigue test [30].

Abo-Qudais and Suleiman [1] adopted the indirect tensile fatigue test by applying a cyclic constant load of 0.1s followed by a 0.4s rest period. When the stiffness was reduced by a quarter of the initial stiffness, samples were subjected to rest periods of 1 day and 3 h, 3 days, 7 days, and 14 days at five different temperatures (0, 15, 30, 45 and 60 °C), and retested again by the indirect tensile fatigue test. These specific test protocols are usually combined with a specific method to calculate the healing factor.

It can be claimed that the intermittent healing test resembles loading in practice closest. This loading application is thought to simulate better the intermittent nature of passing vehicles' axles. In reality, there is always a rest period between the loading of one axle to the next one and, of course, rest in between each truck passage. Unlike in intermittent testing, rest periods will not be constant but will vary depending on the moment in the day and days of the week. In practice, the load levels from trucks will vary, whereas they remain constant in laboratory conditions.

The healing measured using the storage healing method is not directly related to the load pattern in practice. It is sometimes suggested that this type of healing is representative of healing that occurs during the night when the loading frequency is reduced. However, more importantly, storage healing is assumed to correlate with healing that occurs in shorter rest periods and can distinguish between well and poor healing asphalts.

5 Healing Parameters

The analysis and modeling of data gathered by the testing configurations reported above comprise multiple data post-processing procedures based on different theoretical principles that practically determine the success of the methods.

From a general point of view, healing is defined as the measured performance of a damaged specimen after a healing period (f_{healed}), divided over the initial performance ($f_{initial}$), which results in a healing ratio (Hr) presented as Eq. (1).

$$Hr = \frac{f_{healed}}{f_{initial}} \tag{1}$$

When this formula is used, the healing ratio ranges from 0 to 1, where a healing ratio of 0 indicates no recovery, and a healing ratio of 1 indicates full recovery. Different material parameters can be evaluated to establish the restored performance. This strongly depends on the variables associated with the adopted testing methods. In most cases, stiffness-related parameters (S) are considered before and after the rest period. These can also be measured during the rest, thus tracking the evolution of the material restoration [55]. The same formula can be used to establish the restoration of the number of loads to failure (N). In this respect, it should be mentioned that most studies show a higher healing ratio when the stiffness is evaluated, compared to the number of loads to failure [36, 41, 46]. To address these shortcomings, many other techniques have been proposed to analyse the data of the healing tests based on energy ratios (dissipated energy (DE), dissipated energy ratio (DER), the ratio of dissipated energy change (RDEC)) [18, 28] or pseudo stiffness and internal state variables (S_d) [5, 36]. Such parameters can be adopted in fatigue-based tests such as dynamic shear tests in the context of asphalt binders, mastics and mortars. By specifically referring to two-piece asphalt specimens used in fracture-based dynamic shear tests, healing is assessed using the intrinsic healing function (R), which is related to the strength gain due to interfacial cohesion and interdiffusion of molecules at the microcrack faces. For asphalt mixtures, such parameters are linked to uniaxial tension–compression tests, three-point bending and four-point bending tests. In fracture-based approaches, the evaluation criteria adopted in the case of notched three-point bending tests include the peak load at fracture (F_p). Three and four-point bending tests are also implemented to determine the material flexural strength or bend strength (S_h) of a sample, representing the highest stress experienced within the material at its moment of rupture. In the case of semi-circular bending tests, the evaluation of healing includes multiple parameters: the peak load (F_p of the load–displacement curve; the fracture energy (E, which is the energy required to disrupt the interfacial adhesive and cohesive forces; the fracture toughness (FT, describing the material resistance to the propagation of the crack; the secant modulus (MS, dependent on the load–displacement curve; the deformation growth rate (D and the tensile strain growth rate (ε_r) of the material under load.

The peak load (F_p) of the load–displacement curve is also considered when considering the local fracture test of asphalt binders. In the case of the binder bonding strength test implemented at the scale of observation of binders, mastics, and mortars, the pull-off tensile strength (BS) is adopted for the healing assessment. When ductility tests are implemented, the length of break (L) is the considered parameter, which deals with the property of the binder undergoing plastic deformation when a pull force is applied without breaking. In the case of healing of asphalt binders and mastics based on a direct tension test, the fracture strength (S_h) of the original sample is compared to that of the re-fractured sample.

When evaluating the performance after a rest time, it should be kept in mind that it is also possible that the specimen gains strength or stiffness due to the aging process, which is taking place in the sample simultaneously. A test setup should be designed to correct this effect.

Regarding the degree of healing, storage healing tests using fatigue damage often don't report full restoration of the original mixture properties [19] unless rest periods are weeks or longer [12]. However, factors that improve the level of healing have been identified, such as an increased temperature during rest periods [9, 24].

In intermittent healing tests, the healing factor is usually defined as the ratio between the number of loads to failure with and without rest periods (N). If the healing factor is equal to 1, this indicates that the number of loads to failure has not increased by introducing rest periods, implying that no healing has occurred. As such a healing factor from this type of test cannot be compared to a healing factor from a storage healing test, as mentioned in Eq. (1), as a healing ratio of 1 will indicate full healing. Such a healing factor based on N can translate the result of a continuous fatigue test to a test with healing periods. These types of healing observations are often used to determine a healing factor that can be used in a pavement design procedure to translate the result from a continuous fatigue test in the laboratory to a performance in practice [67]. Such a factor can be used in dynamic shear tests and uniaxial Tension–Compression tests. Moreover, a hybrid healing test, combining both storage and intermittent healing approaches, can adopt the parameter N. In the indirect tensile test, the healing assessment can be defined based on the resilient modulus (MR), which is defined as the ratio of the applied stress to the recoverable strain during the application of repeated loads. As a non-destructive method, the ultrasonic pulse velocity (UPV) can be used to evaluate the material strength by measuring the velocity of an ultrasonic pulse passing through the mixture [7, 78]. In the case of the four-point bending fatigue test procedure, the ratio of dissipated energy change (RDEC) has been adopted. This is defined as the difference in dissipated energy between two successive loading cycles divided by the total number of cycles. The RDEC value at the 50% stiffness reduction point, referred to as plateau value (PV), is also considered a healing parameter.

6 Summary

Table 1 summarizes some of the studies carried out on the healing performances of asphalt binders (represented in italic), mastics (represented in bold), mortars (represented in bolditalic) and mixtures (represented in unbold roman, regular). Each study is characterized by the specific testing configurations schematically presented in Figs. 1, 2, 3, and 4, in the case of binder, mastic, mortar and mixture, respectively. Based on the scale of observations and the testing configuration, the healing potential (cohesive or adhesive) can be assessed. The testing configurations also determine the type of approach (fatigue-based or fracture-based) and the healing category (storage, intermittent or hybrid). Finally, based on the post-processing of data, the healing performance is defined through specific healing parameters.

Table 1 Research studies carried out on asphalt binder (represented in italic), mastic (represented in bold), mortar (represented in bolditalic) and mixture (represented in unbold roman, regular)

Authors	Healing mechanism	Damage approach	Testing configuration	Healing category	Healing parameter
Hammoum et al. [26]	*Cohesive*	*Fracture*	*(c)*	*Storage*	F_p
Lu et al. [37]	*Cohesive*	*Fatigue*	*(a)*	*Storage*	*S, N*
Lu et al. [37]	*Cohesive*	*Fatigue*	*(a)*	*Intermittent*	*N*
Bodin et al. [15]	*Cohesive*	*Fatigue*	*(a)*	*Storage*	*S*
Bhasin et al. [14]	*Cohesive*	*Fracture*	*(a)*	*Storage*	*R*
Shen and Sutharsan [58]	*Cohesive*	*Fatigue*	*(a)*	*Intermittent*	*RDEC*
Qiu et al. [48]	*Cohesive*	*Fatigue*	*(a)*	*Storage*	*S*
	Cohesive	*Fracture*	*(b)*	*Storage*	S_h
	Cohesive	*Fracture*	*(d)*	*Storage*	*L*
Tan et al. [65]	*Cohesive*	*Fatigue*	*(a)*	*Storage*	*S*
Santagata et al. [55]	*Cohesive*	*Fatigue*	*(a)*	*Storage*	*S, N*
Canestrari et al. [17]	*Cohesive*	*Fatigue*	*(a)*	*Storage*	*N*
Santagata et al. [54]	*Cohesive*	*Fatigue*	*(a)*	*Storage*	*N*
Pan et al. [43]	*Cohesive*	*Fracture*	*(b)*	*Storage*	F_p, *S, E*
Wang et al. [73]	*Cohesive*	*Fatigue*	*(a)*	*Storage*	S_d
Zhang et al. [76]	*Cohesive*	*Fatigue*	*(a)*	*Storage*	*S, N, DE*
	Cohesive	*Fracture*	*(d)*	*Storage*	*L*
Miglietta et al. [41]	*Cohesive*	*Fatigue*	*(a)*	*Storage*	*S, N*
Aurilio [7]	*Cohesive*	*Fracture*	*(a)*	*Storage*	*E*
Qiu et al. [50]	**Cohesive and adhesive**	**Fracture**	**(g)**	**Storage**	**Sh**
Qiu et al. [49]	**Cohesive and adhesive**	**Fracture**	**(g)**	**Storage**	**Fp**
Santagata et al. [56]	**Cohesive and adhesive**	**Fatigue**	**(e)**	**Storage**	**DE**
Huang et al. [27]	**Cohesive and adhesive**	**Fracture**	**(f)**	**Storage**	**BS**
Lv et al. [39]	**Adhesive**	**Fracture**	**(f)**	**Storage**	**BS**
Al-Mansoori et al. [3]	**Cohesive and adhesive**	**Fatigue**	**(h)**	**Storage**	**F**

(continued)

Table 1 (continued)

Authors	Healing mechanism	Damage approach	Testing configuration	Healing category	Healing parameter
Lv et al. [38]	**Adhesive**	**Fracture**	**(f)**	**Storage**	**BS**
Wang et al. [72]	**Adhesive**	**Fracture**	**(f)**	**Storage**	**BS**
Shi et al. [60]	**Cohesive and adhesive**	**Fatigue**	**(e)**	**Intermittent**	**S**
Van den Bergh and Van de Ven *[68]*	***Cohesive and adhesive***	***Fatigue***	***(i)***	***Hybrid***	***N***
Apostolidis et al. *[4]*	***Cohesive and adhesive***	***Fracture***	***(l)***	***Storage***	***F***
Pan et al. *[43]*	***Cohesive and adhesive***	***Fracture***	***(k)***	***Storage***	F_p, ***S, E***
Grossegger and Garcia [25]	***Cohesive and adhesive***	***Fracture***	***(l)***	***Storage***	***F***
Zhang et al. [77]	***Cohesive and adhesive***	***Fracture***	***(j)***	***Storage***	***BS***
Breysse et al. [16]	Cohesive and adhesive	Fatigue	(p)	Intermittent	S
Abo-Qudais and Suleiman [1]	Cohesive and adhesive	Fatigue	(n)	Hybrid	UPV
Carpenter and Shen [18]	Cohesive and adhesive	Fatigue	(r)	Intermittent	RDEC, PV
Cheng et al. [20]	Cohesive and adhesive	Fatigue	(n)	Storage	MR
Riara et al. [52]	Cohesive and adhesive	Fracture	(o)	Storage	F_p, FT, S, E
Aguirre et al. [2]	Cohesive and adhesive	Fatigue	(q)	Storage	S_h
Jiang et al. [29]	Cohesive and adhesive	Fracture	(o)	Storage	N, MS, RDEC, D, ε_r
Sun et al. [63]	Cohesive and adhesive	Fatigue	(n)	Intermittent	MR
Xiang et al. [70]	Cohesive and adhesive	Fatigue	(r)	Storage	S, N
Ashouri et al. [5]	Cohesive and adhesive	Fatigue	(m)	Intermittent	S_d
Aurilio [7]	Cohesive and adhesive	Fatigue	(m)	Intermittent	S_d, N, UPV

(continued)

Table 1 (continued)

Authors	Healing mechanism	Damage approach	Testing configuration	Healing category	Healing parameter
Zhao [78]	Cohesive and adhesive	Fatigue	(m)	Intermittent	S, UPV
Zhao [78]	Cohesive and adhesive	Fatigue	(n)	Intermittent	S, UPV

BS pull-off tensile strength, *DE* dissipated energy, DER dissipated energy ratio, *D* deformation, *E* fracture energy, *F* fracture force, F_p peak load, *FT* fracture toughness, *L* length at break, *MR* resilient modulus, *MS* secant modulus, *N* number of loading cycles, *PV* plateau value, *R* intrinsic healing, *RDEC* ratio of dissipated energy change, *S* stiffness-related parameter, S_d internal damage state variable, S_h strength, *UPV* ultrasound pulse velocity, ε_r tensile strain

References

1. Abo-Qudais, S., & Suleiman, A. (2005). Monitoring fatigue damage and crack healing by ultrasound wave velocity. *Nondestructive Testing and Evaluation, 20*(2), 125–145.
2. Aguirre, M. A., Hassan, M. M., Shirzad, S., Cooper, S., Jr., Mohammad, L. N., & Negulescu, I. I. (2019). Laboratory testing of self-healing fibers in asphalt mixtures prepared with recycled materials. *Transportation Research Record, 2673*(4), 513–523.
3. Al-Mansoori, T., Norambuena-Contreras, J., Micaelo, R., & Garcia, A. (2018). Self-healing of asphalt mastic by the action of polymeric capsules containing rejuvenators. *Construction and Building Materials, 161*, 330–339.
4. Apostolidis, P., Liu, X., Scarpas, A., Kasbergen, C., & Van De Ven, M. F. C. (2016). Advanced evaluation of asphalt mortar for induction healing purposes. *Construction and Building Materials, 126*, 9–25.
5. Ashouri, M., Wang, Y., Choi, Y. T., & Kim, Y. (2021). Development of healing model and simplified characterization test procedure for asphalt concrete. *Construction and Building Materials, 271*, 121515.
6. Aurilio, R. M., Aurilio, M., & Baaj, H. (2021). The effect of a chemical warm mix additive on the self-healing capability of bitumen. *Journal of Testing and Evaluation, 50*(2).
7. Aurilio, R. M. (2023). *Investigating and developing fatigue-healing characterization of asphalt materials* [Ph.D. Thesis]. University of Waterloo.
8. Ayar, P., Moreno-Navarro, F., & Rubio-Gámez, M. C. (2016). The healing capability of asphalt pavements: A state of the art review. *Journal of Cleaner Production, 113*, 28–40.
9. Baaj, H. (2002). *Comportement à la fatigue des matériaux granulaires traités aux liens hydrocarbons*. INSA.
10. Bahadori, A. M., Mansourkhaki, A., & Ameri, M. A. (2014). Phenomenological fatigue performance model of asphalt mixtures based on fracture energy density. *Journal of Testing and Evaluation, 43*(1), 133–139.
11. Bahia, H. U., Zhai, H., Bonnetti, K., & Kose, S. (1999). Non-linear viscoelastic and fatigue properties of asphalt binders. *J Assoc Asph Pav Tech, 68*, 1–34.
12. Bazin, P., & Saunier, J. (1967). Deformability, fatigue and healing properties of asphalt mixes. In *International conference on the structural design of asphalt pavements.*
13. Bhasin, A., & Little, D. N. (2009). Application of microcalorimeter to characterize adhesion between asphalt binders and aggregates. *Journal of Materials in Civil Engineering, 21*(6), 235–243.
14. Bhasin, A., Little, D. N., Bommavaram, R., & Vasconcelos, K. (2008). A framework to quantify the effect of healing in bituminous materials using material properties. *Road Materials and Pavement Design, 9*(sup1), 219–242.

15. Bodin, D., Soenen, H., & de La Roche, C. (2004). Temperature effects in binder fatigue and healing tests. In *Eurasphalt & Eurobitume Congress, Vienna* (Vol. 82).
16. Breysse, D., De La Roche, C., Domec, V., & Chauvin, J. J. (2003). Influence of rest time on recovery and damage during fatigue tests on bituminous composites. *Materials and Structures, 36*(10), 648–651.
17. Canestrari, F., Virgili, A., Graziani, A., & Stimilli, A. (2015). Modeling and assessment of self-healing and thixotropy properties for modified binders. *International Journal of Fatigue, 70*, 351–360.
18. Carpenter, S., & Shen, S. (2006). Dissipated energy approach to study hot-mix asphalt healing in fatigue. *Trans Res Rec, 1970*, 178–185.
19. Castro, M., & Sánchez, J. A. (2006). Fatigue and healing of asphalt mixtures: Discriminate analysis of fatigue curves. *Journal of Transportation Engineering, 132*, 168–174.
20. Chen, Y., Simms, R., Koh, C., Lopp, G., & Roque, R. (2013). Development of a test method for evaluation and quantification of healing in asphalt mixture. *Road materials and pavement design, 14*(4), 901–920.
21. Dai, Q., Wang, Z., & Mohd Hasan, M. R. (2013). Investigation of induction healing effects on electrically conductive asphalt mastic and asphalt concrete beams through fracture-healing tests. *Construction and Building Materials, 49*, 729e737.
22. Di Benedetto, H., de La Roche, C., Baaj, H., Pronk, A., & Lundström, R. (2004). Fatigue of bituminous mixtures. *Materials and Structures, 37*(3), 202–216.
23. Gong, M., Yang, J., Wei, J., & Zhu, W. (2015). Characterization of adhesion and healing at the interface between asphalt binders and aggregate using atomic force microscopy. *Transportation Research Record, 2506*(1), 100–106.
24. Grant, T. P. (2001). *Determination of asphalt mixture healing rate using the Superpave indirect tensile test*. University of Florida.
25. Grossegger, D., & Garcia, A. (2019). The effect of water and pressure on the self-healing of macro cracks in asphalt mortar beams. *Construction and Building Materials, 229*, 116941.
26. Hammoumm, F., Piau, J., & Stefani, C. (2002). *Experimental investigation of fracture and healing of bitumen at pseudo-contract of two aggregates.*
27. Huang, W., Lv, Q., & Xiao, F. (2016). Investigation of using binder bond strength test to evaluate adhesion and self-healing properties of modified asphalt binders. *Construction and Building Materials, 113*, 49–56.
28. Isailović, I., Wistuba, M. P., & Falchetto, A. C. (2017). Influence of rest period on asphalt recovery considering nonlinearity and self-heating. *Construction and Building Materials, 140*, 321–327.
29. Jiang, J., Ni, F., Wu, F., Sadek, H., & Lv, Q. (2019). Evaluation of the healing potential of asphalt mixtures based on a modified semi-circular bending test. *Construction and Building Materials, 196*, 284–294.
30. Kim, Y. R., Little, D. N., & Lytton, R. L. (2003). Fatigue and healing characterization of asphalt mixtures. *Journal of Materials in Civil Engineering, 15*(1), 75–83.
31. Kim, Y., Allen, D. H., & Little, D. N. (2007). Computational constitutive model for predicting nonlinear viscoelastic damage and fracture failure of asphalt concrete mixtures. *International Journal of Geomechanics, 7*, 102–110.
32. Lee, H. J., Daniel, J. S., & Kim, Y. R. (2000). Continuum damage mechanics-based fatigue model of asphalt concrete. *Journal of Materials in Civil Engineering, 105*, 1005–1112.
33. Leegwater, G. A., Scarpas, A., & Erkens, S. M. J. G. (2018). The influence of boundary conditions on the healing of bitumen. *Road Materials and Pavement Design, 19*, 571–580.
34. Leegwater, G., Taboković, A., Baglieri, O., Hammoum, F., & Baaj, H. (2020). Terms and definitions on crack-healing and restoration of mechanical properties in bituminous materials. In *RILEM international symposium on bituminous materials* (pp. 47–53). Springer.
35. Li, Q., Jong, L. H., & Kim, T. W. (2012). A simple fatigue performance model of asphalt mixtures based on fracture energy. *Construction and Building Materials, 27*(1), 605–611.
36. Little, D. N., Bhasin, A., & Darabi, M. K. (2015). Damage healing in asphalt pavements: Theory, mechanisms, measurement, and modeling. In *Advances in asphalt materials: road and pavement construction.*

37. Lu, X., Soenen, H., & Redelius, P. (2003). Fatigue and healing characteristics of bitumens studied using dynamic shear rheometer. In *6th RILEM Symposium PTEBM* (Vol. 3).
38. Lv, Q., Huang, W., Zheng, M., Hao, G., Yan, C., & Sun, L. (2020). Investigating the asphalt binder/mastic bonding healing behavior using bitumen bonding strength test and X-ray Computed Tomography scan. *Construction and Building Materials, 257*, 119504.
39. Lv, Q., Huang, W., Zhu, X., & Xiao, F. (2017). On the investigation of self-healing behavior of bitumen and its influencing factors. *Materials and Design, 117*, 7–17.
40. Mazzoni, G., Stimilli, A., Cardone, F., & Canestrari, F. (2017). Fatigue, self-healing and thixotropy of bituminous mastics including aged modified bitumens and different filler contents. *Construction and Building Materials, 131*, 496–502.
41. Miglietta, F., Tsantilis, L., Baglieri, O., & Santagata, E. (2021). A new approach for the evaluation of time-temperature superposition effects on the self-healing of bituminous binders. *Construction and Building Materials, 287*, 122987.
42. Nejad, F. M., Hamedi, G. H., & Azarhoosh, A. R. (2013). Use of surface free energy method to evaluate effect of hydrate lime on moisture damage in got mix asphalt. *Journal of Materials in Civil Engineering., 25*(8), 1119–1126.
43. Pan, C., Tang, P., Riara, M., Mo, L., Li, M., & Guo, M. (2018). Effect of healing agents on crack healing of asphalt and asphalt mortar. *Materials (Basel), 11*, 1373.
44. Park, K., & Paulino, G. H. (2013). Cohesive zone models: A critical review of traction–separation relationships across fracture surfaces. *Applied Mechanics Reviews, 64*, 1–20.
45. Pronk, A. C.: Partial healing—a new approach for the damage process during fatigue testing of asphalt specimen. In *Asphalt concrete: simulation, modeling, and experimental characterization* (2006).
46. Pronk, A. C., & Cocurullo, A. (2009). Investigation of the PH model as a prediction tool in fatigue bending tests with rest periods. In *Advanced testing and characterisation of bituminous materials*. Loizos, Partl, Taylor & Francis Group.
47. Qiu, X., Cheng, W., Xu, W., Xiao, S., & Yang, Q. (2020). Fatigue evolution characteristic and self-healing behaviour of asphalt binders. *International Journal of Pavement Engineering*, 1–12.
48. Qiu, J., Van de Ven, M. F. C., Wu, S., Yu, J., & Molenaar, A. A. A. (2009). Investigating the self healing capability of bituminous binders. *Road Materials and Pavement Design, 10*(sup1), 81–94.
49. Qiu, J., Van de Ven, M. F., Wu, S., Molenaar, A. A., & Yu, J. (2014). Self-healing characteristics of bituminous mastics using a modified direct tension test. *Journal of Intelligent Material Systems and Structures, 25*, 58–66.
50. Qiu, J., Van de Ven, M., Wu, S., Yu, J., & Molenaar, A. (2012). Evaluating self-healing capability of bituminous mastics. *Experimental Mechanics, 52*(8), 1163–1171.
51. Radeef, H. R., Abdul Hassan, N., Zainal Abidin, A. R., Mahmud, M. Z. H., Mohd Satar, M. K. I., Mohd Warid, M. N., & Al Saffar, Z. H. (2021). Determining fracture energy in asphalt mixture: A review. *Conference Series: Earth and Environmental Science, 682*(1), 12069 (2021).
52. Riara, M., Tang, P., Mo, L., Javilla, B., Chen, M., & Wu, S. (2018). Systematic evaluation of fracture-based healing indexes of asphalt mixtures. *Journal of Materials in Civil Engineering, 30*(10), 04018264.
53. Rondón, H. A., Zafra, C. A., & Reyes, F. A. (2021). Fatigue in asphalt mixtures—A summary to understand the complexity of its mathematical modeling. *Journal of Physics: Conference Series, 2118*, 012009.
54. Santagata, E., Baglieri, O., Dalmazzo, D., & Tsantilis, L. (2017). Investigating cohesive healing of asphalt binders by means of a dissipated energy approach. *International Journal of Pavement Research and Technology, 10*(5), 403–409.
55. Santagata, E., Baglieri, O., Tsantilis, L., & Dalmazzo, D. (2013). Evaluation of self-healing properties of bituminous binders taking into account steric hardening effects. *Construction and Building Materials, 41*, 60–67.
56. Santagata, E., Baglieri, O., Tsantilis, L., Dalmazzo, D., & Chiappinelli, G. (2016). Fatigue and healing properties of bituminous mastics reinforced with nano-sized additives. *Mechanics of Time-Dependent Materials, 20*(3), 367–387.

57. Shen, S., Chiu, H. M., & Huang, H. (2010). Characterization of fatigue and healing in asphalt binders. *Journal of Materials in Civil Engineering, 22*(9), 846–852.
58. Shen, S., & Sutharsan, T. (2011). Quantification of cohesive healing of asphalt binder and its impact factors based on dissipated energy analysis. *Road Materials and Pavement Design, 12*(3), 525–546.
59. Saha, G., & Biligiri, K. P. (2015). Fracture damage evaluation of asphalt mixtures using Semi-Circular Bending test based on fracture energy approach. *Engineering Fracture Mechanics, 142*, 154–169.
60. Shi, C., Luo, R., Liu, H., & Yang, J. (2021). Evaluation of self-healing characteristics of asphalt materials by intermittent loading tests. *Road Materials and Pavement Design.*
61. Song, I., Little, D. N., Masad, E. A., & Lytton, R. L. (2005). Comprehensive evaluation of damage in asphalt mastics using X-ray CT, continuum mechanics, and micromechanics (with discussion). *Asphalt Paving Technology, 74.*
62. Sun, G., Hu, M., Sun, D., Deng, Y., Ma, J., & Lu, T. (2020). Temperature induced self-healing capability transition phenomenon of bitumens. *Fuel, 263*, 116698.
63. Sun, G., Sun, D., Guarin, A., Ma, J., Chen, F., & Ghafooriroozbahany, E. (2019). Low temperature self-healing character of asphalt mixtures under different fatigue damage degrees. *Construction and Building Materials, 223*, 870–882.
64. Sun, D., Sun, G., Zhu, X., Guarin, A., Li, B., Dai, Z., & Ling, J. (2018). A comprehensive review on self-healing of asphalt materials: Mechanism, model, characterization and enhancement. *Advances in Colloid and Interface Science, 256*, 65–93.
65. Tan, Y., Shan, L., Kim, Y. R., & Underwood, B. S. (2012). Healing characteristics of asphalt binder. *Construction and Building Materials, 27*(1), 570–577.
66. Underwood, B. S., Kim, Y. R., & Guddati, M. N. (2010). Improved calculation method of damage parameter in viscoelastic continuum damage model. *International Journal of Pavement Engineering, 11*(6), 459–476.
67. Van den Bergh, W. (2011). *The effect of ageing on the fatigue and healing properties of bituminous mortars.*
68. Van den Bergh, W., & Van de Ven, M. F. C. (2012). The influence of ageing on the fatigue and healing properties of bituminous mortars. *Procedia-Social and Behavioral Sciences, 53*, 256–65.
69. Varma, R., Balieu, R., & Kringos, N. (2021). A state-of-the-art review on self-healing in asphalt materials: Mechanical testing and analysis approaches. *Construction and Building Materials, 310*, 125197.
70. Xiang, H., Zhang, W., Liu, P., & He, Z. (2020). Fatigue-healing performance evaluation of asphalt mixture using four-point bending test. *Materials and Structures, 53*, 1–10.
71. Xie, W., Castorena, C., Wang, C., & Kim, Y. R. (2017). A framework to characterize the healing potential of asphalt binder using the linear amplitude sweep test. *Construction and Building Materials, 154*, 771–779.
72. Wang, C., Chen, Y., & Gong, G. (2021). Cohesive and adhesive healing evaluation of asphalt binders by means of the LASH and BBSH tests. *Construction and Building Materials, 282*, 122684.
73. Wang, C., Xie, W., & Underwood, B. S. (2018). Fatigue and healing performance assessment of asphalt binder from rheological and chemical characteristics. *Materials and Structures, 51*(6), 1–12.
74. Williams, D., Little, D. N., Lytton, R. L., Kim, Y. R., & Kim, Y. (2001). *Microdamage healing in asphalt and asphalt concrete, volume II: Laboratory and field testing to assess and evaluate microdamage and microdamage healing.* No. FHWA-RD-98-142.
75. Wool, R. P., & O'Connor, K. M. (1981). A theory crack healing in polymers. *Journal of Applied Physics, 52*, 5953–5963.
76. Zhang, H., Bai, Y., & Cheng, F. (2018). Rheological and self-healing properties of asphalt binder containing microcapsules. *Construction and Building Materials, 187*, 138–148.
77. Zhang, H., Li, H., Abdelhady, A., Mo, D., Yang, B., & Harvey, J. (2019). Optimum filler–bitumen ratio of asphalt mortar considering self-healing property. *Journal of Materials in Civil Engineering, 31*(8), 04019166.

78. Zhao, D. (2023). *Development and evaluation of testing protocols for fatigue damage and crack-healing of bituminous mixtures [Ph.D. Thesis]*. University of Waterloo.

Modeling the Crack-Healing Behavior of Asphalt Pavement Materials

Salvatore Mangiafico, Yasmina Mahmoudi, Fabrizio Miglietta, Roberto M. Aurilio, F. Williams, Cédric Sauzéat, Orazio Baglieri, and Hassan Baaj

Abstract Various models have been developed to simulate the self-healing properties of asphalt materials. These models are based on experimental observations and numerical simulations, aiming to describe the healing behaviour of materials under different thermal and loading conditions. Since the processes governing self-healing can be observed at different scales (nano-/micro-, intermediate-, and macro-) , existing modelling approaches can be classified based on the scales of observation. This section presents a literature review of diverse models developed and proposed to analyze and estimate self-healing in asphalt materials. It must be highlighted that modeling of self-healing bituminous materials shows room for improvement, mainly due to the complex, multi-scale nature of the healing processes and the lack of a unified approach in existing research.

1 Introduction

Among the existing literature on self-healing asphalt materials, only a relatively small fraction of the published studies focused on the modeling aspect, from rheological models to phenomenological regressions of experimental data to numerical simulations at the material and structure scale. Unlike other, more mature research

S. Mangiafico (✉) · Y. Mahmoudi · F. Williams · C. Sauzéat
ENTPE, Ecole Centrale de Lyon, CNRS, LTDS, UMR 5513, Vaulx-en-Velin, France
e-mail: salvatore.mangiafico@entpe.fr

Y. Mahmoudi
Eiffage Infrastructures, Corbas, France

F. Miglietta · O. Baglieri
Department of Environment, Land and Infrastructure Engineering, Politecnico di Torino, 24, c.so Duca degli Abruzzi, 10129 Turin, Italy

R. M. Aurilio · H. Baaj
Department of Civil and Environmental Engineering, University of Waterloo, 200 University Ave W, N2L 3G1, Waterloo, ON, Canada

© The Author(s), under exclusive license to Springer Nature Switzerland AG 2025

H. Baaj and O. Baglieri (eds.), *Crack-Healing of Asphalt Pavement Materials*, RILEM State-of-the-Art Reports 45, https://doi.org/10.1007/978-3-031-99443-2_5

subjects in the domain of pavement materials, it is difficult to identify in the dedicated literature, a clear axis of investigation based on solid approaches shared by a large part of the scientific community. This sporadic research reflects the lack of a precise definition of the observed phenomena behind the restoration of material properties after damage and an insufficient understanding of their fundamental underlying mechanisms [1–3].

The literature shows research works on different material scales, from binders to mastics and mixtures. However, a comprehensive multiscale approach, establishing quantitative relationships between experimental observations, even those limited to a laboratory scale, is still missing. In fact, state-of-the-art knowledge on the healing capability of asphalt materials is still mainly related to experimental mechanical characterization of the phenomenon, often qualitatively but more rarely qualitatively. Even when a "healing index" is proposed and calculated (based on various mechanical properties), the outcome strongly depends on the mechanical tests' technique and conditions. For these reasons, among others, healing as a truly definable material property is still far from a comprehensive model.

The previous chapters of this report highlight the need for a proper definition of the different complex phenomena related to damage occurrence and restoration prior to approaching a quantification of the self-healing properties of asphalt materials. Furthermore, according to the test methods considered (and therefore the mechanical properties used to assess restoration), comparisons are made between different domains of the behavior of the materials (typically, small strain and large strain behavior). It is not uncommon that observations regarding restoration of stiffness and/or strength are alternatively looked at or compared. However, these properties are not directly correlated and reflect very different aspects of the thermomechanical behavior of the materials.

Healing properties for asphalt mixtures refer to their ability to self-repair and recover from damage, particularly microcracks that, over time, can develop into macrocracks due to traffic and climatic conditions, ultimately leading to significant deterioration [4–6]. The mechanism of healing remains unclear, but according to some researchers, it consists of a three-step process [7, 8]. This process implies that micro-cracks, or at least minor damages, occur in asphalt materials due to traffic or other loads as well as temperature changes.

The first stage corresponds to the transport of asphalt binder, also known as asphalt binder flow. Asphalt binders can migrate toward the damaged area under certain conditions depending on temperature, pressure, and moisture. The second stage is called wetting, which consists of the filling of micro-cracks with asphalt binder, allowing them to close. The surfaces of cracks adhere to each other and are driven by surface energy. Thus, a gain of coherence occurs. This stage depends on multiple internal factors, such as binder characteristics and the composition of mixtures, as well as other external factors, such as temperature and climatic conditions in general. The last stage is the complete recovery of properties, allowed by diffusion and randomization of molecules. The healing process allows the restoration of the structural integrity of the mixtures and prevents any possible damage propagation.

The first and second stages are considered short-term processes, whereas the last stage is regarded as a long-term process [9]. Various models have been developed to predict the healing properties of asphalt materials based on experimental observations and numerical simulations, aiming to simulate and predict the healing behavior under different conditions. As previously discussed, researchers concur that the self-healing phenomenon encompasses three phases (flow, wetting, and recovery). Therefore, the processes governing self-healing can be observed at different scales: nano-/micro-, intermediate, and macro-. This section presents a comprehensive literature review of diverse models developed to enhance the comprehension and prediction of self-healing in asphalt materials, classified based on the scales of observation. The approach based on ViscoElastic Continuum Damage (VECD) theory-based modeling is also considered. According to its general principles and use in the literature, this approach spans the different scales of the materials, allowing fatigue and healing properties of binders, mastics, and mixtures to be modelled.

2 Healing Models Based on Molecular Diffusion Theories

Laboratory-based test methods used to quantify healing typically involve damaging the sample in a fatigue or partial fracture mode and then allowing it to recover over time. While this approach effectively differentiates healing characteristics among different asphalt binders or asphalt mixtures, it has a drawback. The magnitude of healing is dependent on the test method, making it challenging to model mixture behavior under various laboratory or field conditions. However, this limitation can be overcome by using an analytical solution based on intrinsic material properties related to the healing process. By incorporating these properties, one can quantitatively model and predict healing observed in laboratory fatigue tests or field conditions [6].

Most models of phenomena occurring during the second and third steps of the healing process (wetting and interdiffusion of asphalt binder) are inspired by the behavior of long polymer chains. When the material experiences damage, such as cracks or micro-cracks, the polymer chains can move and flow to fill the voids created by the defects. De Gennes applied the Theory of Reptation to polymers in the context of self-healing in asphalt materials [10]. Reptation is a fundamental concept in polymer physics, describing the slow and random movement of molecular chains when they are entangled with each other. In asphalt materials, the asphalt binder may be represented by long, closely entangled polymer chains, forming a complex network. When the material experiences damage, such as cracks or micro-cracks, the polymer chains can be partially released from their entanglement and mutual immobilization. Due to their molecular mobility, these polymer chains can then begin to move slowly and randomly. This can cause them to move towards the damaged areas of the material, where they have more space to maneuver. They can then intercalate between the defects and re-entangle, contributing to the self-healing process.

The self-healing process is thus mainly governed by the mobility and diffusion of the polymer chains within the material. By moving in a disordered manner, the chains can rearrange themselves to fill the voids created by the cracks and repair the damage incurred by the material. De Gennes' approach highlights the importance of understanding the molecular mobility of polymer chains to grasp better self-healing in asphalt materials [10].

In addition, Schapery introduced a correlation between the work of cohesion and material properties associated with the wetting of crack surfaces [11, 12]. Using this correlation, the wetting distribution function or rate of wetting for a crack surface can be expressed as follows in Eq. (1):

$$\frac{d\phi(t, X)}{dt} = \dot{a}_b = \beta\left(\frac{1}{D_1 k_m}\left\{\frac{\pi W_c}{4(1-\nu^2)\sigma_b^2\beta} - D_0\right\}\right)^{-\frac{1}{m}} \quad (1)$$

where W_c is the work of cohesion, ν is the Poisson's ratio, D_0, D_1 and m are creep compliance parameters (obtained by fitting $D(t) = D_0 + D_1.t^m$), and k_m is a material constant.

Parameters $\dot{a}_b$, β and σ_b are presented in Fig. 1. β is the length of the crack or healing process zone. The tensile stress between the crack surfaces within this healing process zone is denoted by σ_b. The advancement rate of the crack tip within the healing process zone, resulting in wetting between the crack surfaces, is represented by $\dot{a}_b$.

Wool and O'Connor considered a molecular approach with five distinct phases: surface rearrangement, surface approach, wetting, diffusion, and randomization [13]. They developed a model using the same reptation model as the De Gennes model. They then reversed the approach to determine the healing time caused by the wetting and the molecular diffusion between the wetted crack faces. They defined this using a convolution integral of a combination of the intrinsic healing $R_h(t)$ and a wetting distribution function $\phi(t, X)$ as follows in Eq. (2):

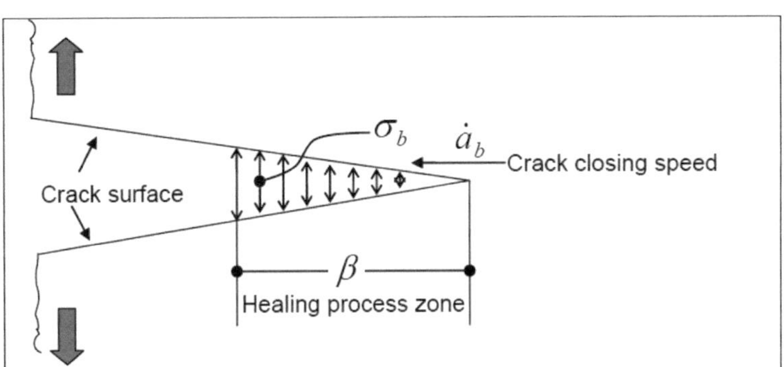

Fig. 1 Crack propagation and fracture process/healing zone in mode I loading [6]

$$R = \int_{\tau=\infty}^{\tau=t} R_h(t-\tau)\frac{d\phi(\tau)}{d\tau}d\tau \quad (2)$$

where R is the net macroscopic healing function representing the net effect of the wetting and strength gain processes related to cohesion and interdiffusion. $R_h(t)$ is the intrinsic healing function that defines instantaneous strength gained by cohesion (R_0) and time-dependent strength gained by interfacial diffusion, $\phi(t, X)$ is the wetting function, τ is the time variable, and t is the present time.

However, models based on analogy polymers may not sufficiently describe the self-healing phenomenon in asphalt materials, given the significant complexity of asphalt binder composition. Therefore, researchers have relied on these theories and enriched them by incorporating additional elements that align with the characteristics of asphalt materials.

Bahsin et al. [6] proposed a healing model that described the time-dependent behaviour of intrinsic healing based on Wool and O'Connor, and Schapery works. They used a simplified form of the intrinsic healing function of asphalt binder, as shown in Eq. (3):

$$R_h(t) = R_0 + p\left(1 - e^{-qt^r}\right) \quad (3)$$

where $R_h(t)$ is a time-dependent function that represents the increase in mechanical property of the wetted crack faces, typically as a fraction or percentage of the same property for the intact material. The term R_0 accounts for the immediate healing of wetted surfaces, primarily driven by interfacial cohesion resulting from surface-free energy. The second part of the equation represents the kinetics of a phase transformation, where molecules migrate and randomize across the interface. The constants q and r are time-independent parameters determined by intrinsic material properties, defined through dynamic shear rheometry (DSR), and influenced by molecular weight and activation energy. The significant interest in this function lies in its versatility, as it can be used to represent any mechanical property of interest. The authors determined the healing function parameters for three different asphalt binder types using Dynamic Shear Rheometer (DSR) tests (Fig. 2) [14]. For this purpose, two asphalt binder specimens, each with a diameter of 28 mm and a thickness of 3.5 mm, were secured to the two end plates of the DSR. These specimens were then brought into contact with each other, and changes in the shear modulus were recorded by subjecting them to small strain cyclic loads at various time intervals. To ensure accurate measurements, the change in the shear modulus was normalized relative to a single intact specimen of equivalent size.

The simulations of the tested asphalt binders closely aligned with the observed increase in fatigue cracking life of the corresponding asphalt mixtures when rest periods were introduced. Furthermore, a more accurate correlation was obtained when considering the wetting distribution in addition to the healing function, as presented previously.

a) Two bitumen specimens affixed to the two end plates of the DSR

b) Bitumen specimen after being brought into contact with each other to measure change in composite G* over time

c) Separation of the test specimens after completion of the test

Fig. 2 Photos of the experimental test methods used by Bhasin et al. to obtain parameters for the healing function using the DSR [6]

Sun et al. developed a Molecular Dynamics (MD) simulation used as a tool to examine the hypothesis of the healing mechanism and evaluate the self-healing capability of neat and SBS-modified asphalt binders [15]. The density and glass transition temperature of the two types of asphalt binders were initially analyzed using molecular dynamics simulation. It was found that the density and temperature of the asphalt models were almost identical to those of real asphalt binders. This means that the MD models and force fields adopted provide sufficient accuracy for predicting the properties of asphalt binders. An artificial crack was created in the MD model to simulate the diffusion of molecules across a crack interface. Three indices calculated from MD simulation results were put forward to evaluate the self-healing capability of asphalt binders, including diffusion coefficient, activation energy, and pre-exponential factor. It suggests that the developed MD methods studying the self-healing property of asphalt binders can provide accurate results.

Zhou et al. applied this approach to evaluate the self-healing mechanism and hydrophobic performance of asphalt modified by siloxane and polyurethane [16], obtaining successful simulations.

3 Healing Models at the Intermediate Scale

Many authors agree that the models presented above are irrelevant for a comprehensive description of the healing phenomenon at scales beyond the nanoscopic or microscopic level. While these models effectively account for the healing of nano-

and micro-cracks, they fall short in explaining the healing of intermediate and macro-cracks, which have been experimentally observed. Consequently, there is a need to explore alternative modelling approaches better suited for larger scales and capable of offering a more accurate representation of the healing phenomenon observed in real-world conditions. This has led researchers to explore other modelling approaches, summarized below.

3.1 Capillary Flow Models

The capillary flow models developed to understand the healing phenomena in asphalt materials are based on the principle of capillary forces. These models take into account the properties of the asphalt binder as a viscous liquid and the interactions between the crack surfaces and the asphalt binder. When cracks form in asphalt materials, the viscous fluid of the asphalt binder can be drawn into the cracks through capillary action. Capillary forces are responsible for the phenomenon where liquids rise or penetrate tiny open spaces, such as microscopic cracks. In the case of asphalt materials, these capillary forces can lead to the flow of asphalt binder into the cracks, allowing it to fill the voids created by the cracks and contribute to self-healing. Capillary flow models examine the geometric characteristics of the cracks, the viscosity of the asphalt binder, and the interactions between the asphalt binder and the crack surfaces. They also consider the physicochemical properties of the asphalt binder and aggregates in the asphalt mixture.

In García et al. [17], healing levels have been determined by comparing the three-point bending strength of semi-circular test samples at − 20 °C before and after the healing process. Various healing levels have been achieved by heating mixture samples at different temperatures. These healing levels were then fitted to the model proposed by García et al. [18–20]. To do so, the authors explained that the healing of open cracks in asphalt mixtures occurs when some points on both faces of a crack come into contact. Capillary phenomena initiate from these contact areas and can extend through the crack if the temperature of the mixture is sufficiently high for the asphalt binder to behave as a Newtonian fluid. During the healing process, asphalt binder can be considered a liquid with density δ surface tension γ and a contact angle θ in the wetting front (Fig. 3).

The capillary phenomena can be studied by considering the equilibrium of forces:

- The weight of the healing material filling the crack (F_h) under gravitational acceleration g
- The hydrostatic force exerted by the weight of asphalt mixtures in the crack area (F_f),
- The dissipation force caused by the movement of asphalt binder against the walls of the crack (F_d),
- And the force exerted by the surface tension of liquid asphalt binder (F_t).

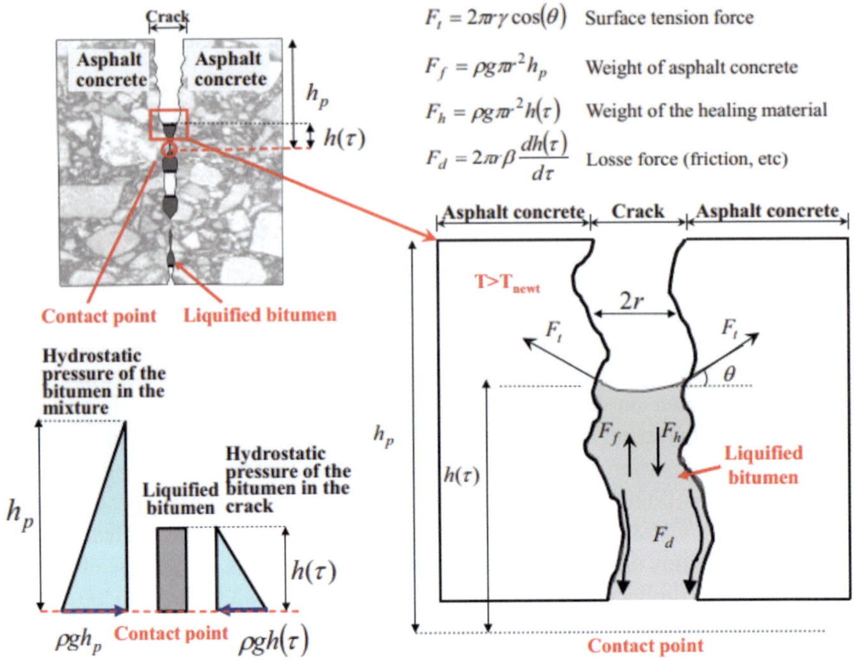

Fig. 3 Representation of the forces in action within a crack in an asphalt concrete during the healing process [17, 21]

According to Garcia et al. [19], it is postulated that a crack with a radius r located at a depth h_p in the mixture, will begin to heal at a constant rate once a certain amount of energy τ has been induced into the pavement. It is further explained that the healed area of the crack $A_{E(\tau)}$, after undergoing heating, considering a specific number of potential contact points n where the healing process may occur, can be expressed as follows in Eqs. (4) using (5):

$$A_{E(\tau)} = C.e^{-D\tau}\left(-1 + e^{\frac{D\tau}{2}}\right) \tag{4}$$

where:

$$C = \pi n\left(h_p + \frac{2\gamma}{\rho g r}\right)^2 \text{and } D = \frac{\rho g r}{\beta} \tag{5}$$

The parameter β is introduced as a dissipation factor to account for potential sources of energy dissipation during the healing process. Furthermore, the maximum force that a asphalt mixture semi-circular test specimen, with maximum strength σ_u, the distance between the supports L, and height H, can resist under three-point bending after some time above the Newtonian temperature, can be expressed as in Eqs. (6) using (7):

$$F_b(\tau) = C_1.e^{-D\tau}\left(-1 + e^{\frac{D\tau}{2}}\right)^2 \tag{6}$$

where:

$$C_1 = 8\frac{\sigma_u.C}{LH} \tag{7}$$

The healing level $S(t)$ can be characterized (experimentally and analytically) as the ratio between the ultimate force (Fi) observed in the test specimens during a three-point bending test and the ultimate force ($F_b(t)$) measured in the semi-circular specimens after the healing process (Eq. 8). Here, the parameter t indicates the amount of energy applied to the asphalt mixtures test sample during the healing process.

$$S(\tau) = \frac{F_b(t)}{F_i} \tag{8}$$

Finally, by substitution, $S(t)$ can be written in Eq. (9):

$$S(\tau) = \frac{C_1}{F_I}e^{-D\tau}\left(-1 + e^{\frac{D\tau}{2}}\right)^2 \tag{9}$$

3.2 Phase Field Healing Model

The principle of phase field healing models in asphalt mixtures is a modelling approach that facilitates the simulation of the self-healing process of cracks and defects in asphalt materials. These models represent the damaged material as having distinct phases or regions with different physical properties. Each phase corresponds to a specific area of the material, such as a cracked zone or an intact zone. Mathematically, these phases are described using a phase field—a continuous variable that takes specific values in each phase throughout the material. The fundamental principle of these models is based on the continuous variation of the phase field, which enables the representation of the healing progression throughout the material. As time passes, the phase field evolves, allowing different areas of the material to transition from a damaged phase to a healed phase [22, 23].

Hence, Kringos et al. introduced a novel finite element-based model that examines the healing phenomenon of asphalt materials from a thermodynamic perspective [23]. This model utilizes the phase field approach previously described to simulate the healing process, incorporating a modified version of the Cahn–Hilliard and Flory–Huggins equations. This model has been integrated into the 3D finite element system CAPA-3D. Atomic Force Microscopy (AFM) investigations revealed heterogeneity in the microstructure of asphalt binder due to its tendency to phase-separate under certain kinetic conditions [24]. Consequently, interfaces between the different phases are present. From a mechanical standpoint, interfaces between materials with different stiffness induce natural stresses that can cause cracks when subjected to mechanical or environmental loads. This process can lead to micro-cracks, which, if left unchecked, may eventually merge and form larger cracks. However, during the stage of micro-crack development, the material's properties can be restored by removing the cracks through changes in the material's kinetics, such as increasing the temperature. This phase rearrangement allows the material to heal and recover its mechanical properties. This phase rearrangement process is modelled by the phase field theory and serves as the basis for a finite element-based healing model for asphalt binder.

From a thermodynamic point of view, macroscopic properties are directly linked to the characteristics of single particles (e.g., atoms, molecules). The description of a system at the macroscopic scale depends on its microscopic characteristics. Therefore, to study the properties of a material undergoing phase changes, it is essential to identify all independent macroscopic characteristics representing the system's state. The thermodynamic state of a system comprises two components: (i) all the external conditions influencing the system and (ii) all the independent properties of the system. When all the independent properties of the system are in the same state when it reaches equilibrium, it is referred to as a homogeneous system.

The Gibbs energy G and the Helmholtz potential F are two thermodynamic quantities that play a central role in the study of systems in equilibrium. These quantities are related to the chemical potentials of different phases of a material and the system's free energy. The Gibbs energy represents the system's total energy available to work at constant pressure and temperature. The Gibbs energy is minimized when the system reaches equilibrium at constant pressure and temperature. Conversely, the Helmholtz potential reflects the total energy available to perform work at constant volume and is minimized when the system reaches equilibrium at constant temperature and volume.

The presence of phases (e,g, α and β) in a system is determined by the Gibbs phase law. At phase equilibrium, when the chemical potential of phase α (μ_α) is equal to that of phase β (μ_b), both phases stably coexist. However, in real materials, the system temperature deviates from the equilibrium temperature, violating the equality of chemical potentials. This deviation creates a driving force for phase separation (Fig. 4). Phase fractions can only be altered under non-equilibrium conditions, where a driving force exists.

In the context of Kringos' phase field healing model, these thermodynamic quantities are used to assess the equilibrium state of different phases present in the material. When the material undergoes damage, it generally exists in a non-equilibrium state. In this state, the chemical potentials of different phases are imbalanced, and the system faces an inequality in free energy between the damaged and healed phases. Temperature plays a crucial role in the healing process. An increase in temperature accelerates chemical reactions and diffusion processes within the material, enhancing the mobility of atoms or molecules. This allows damaged phases to reorganize more rapidly into a healed configuration. Thus, temperature influences the speed of the healing process.

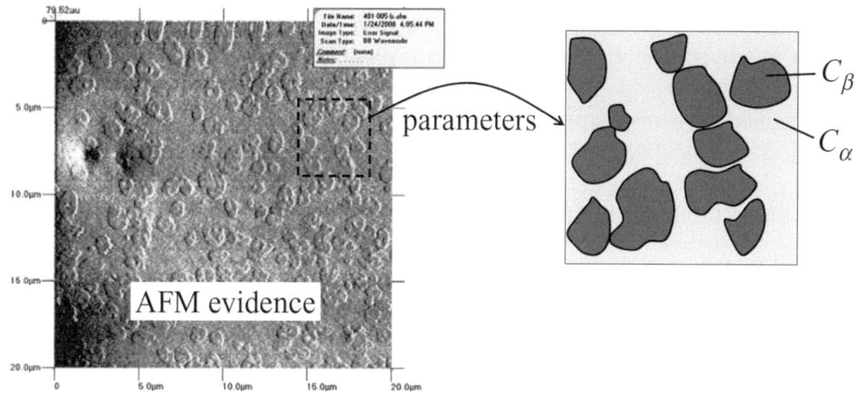

Fig. 4 AFM evidence of phase-separation in asphalt binder [23, 24]

In Kringos' phase field healing model, the evolution of the continuous phase field φ(x, t) is governed by a phase field equation that considers local variations in free energy due to phase variations. By minimizing the free energy of the system, the phase field evolves to reach a new equilibrium state where mechanical properties are restored. Variations in temperature and differences in free energy between damaged and healed phases influence the direction and speed of this healing process.

4 Models for Macroscopic Scale Healing

As previously mentioned, mechanical test-based modeling is heavily influenced by the specific types of tests conducted and their conditions. Factors such as the imposed stress amplitudes, the durations and frequencies of rest periods, and the types of applied loading can significantly impact the outcomes of the modelling process. It is important to acknowledge that the choice of test parameters plays a critical role in determining the accuracy and reliability of the model. Different tests may capture specific aspects of the material's behavior, and combining the results from multiple tests can provide a more comprehensive understanding of the self-healing phenomenon.

Despite the dependency on test conditions affecting their quantitative accuracy, such models can still yield valuable qualitative results. Thus, while acknowledging their limitations, these models remain important tools for studying self-healing materials. It is worth mentioning that, throughout this section, we have respected the terminology used by the original authors for self-healing and the associate phenomena. The TC-278 CHA (Leegwater et al. 2022), proposed specific terms and definitions for self-healing, recovery, and restoration that are summarized in Sect. 8 in Chapter "Intrinsic Self-Healing of Asphalt Pavements" of this book.

One of the primary challenges in modelling self-healing lies in distinguishing it as a distinct recovery phenomenon from other reversible effects, such as thixotropy and heating. Canestrari et al. proposed a new analytical model to simulate experimental data obtained from time sweep tests on asphalt binders with various rest periods [25]. These tests involved evaluating multiple binders using a Dynamic Shear Rheometer (DSR) with a plate-plate configuration (diameter: 8 mm, gap: 2 mm). Time sweep tests with multiple rest periods were conducted, involving a series of loading and rest phases.

Regarding the analytical analysis part, they assumed that the total number of cycles resisted by the material tested after n rest period (N_{tot}) is equal to (Eq. 10):

$$N_{tot} = N_0 + \sum_{i=1}^{n} \Delta N(i) \tag{10}$$

where N_0 is related to the initial material's strength and $\Delta N(i)$ is the number of recovered cycles (i.e. the number of cycles needed to achieve 65% of $G^{*} sinsin(\delta)_0$) after each rest period. They show that N_{tot} could also be written as follows in Eq. 11:

$$\begin{aligned} N_{tot} &= N_0 + \sum_{i=1}^{n} \Delta N_H(i) + \sum_{i=1}^{n} \Delta N_\infty \\ &= N_0 + N_H(n) + N_T(n) \end{aligned} \tag{11}$$

where $N_H(n)$ can be attributed to the self-healing effect and $N_T(n)$ is due to thixotropy. They used a simple model to represent the self-healing contribution as presented in Eq. (12):

$$\Delta N_H(i) = k_1 e^{-\alpha i} \tag{12}$$

where k_1 and α are parameters, whose value depends on the material. Thus, $N_H(n)$ can be expressed as follows in Eq. (13):

$$N_H(n) = \sum_{i=1}^{n} k_1 e^{-\alpha i} = k_1 \frac{1 - e^{-\alpha n}}{e^{\alpha} - 1} \tag{13}$$

Hence, after a virtually infinite number of rest periods, the cumulative self-healing contribution N_{fH} converge to the expression (Eq. 14):

$$N_{fH} = N_H(n) = \frac{k1}{e^{\alpha} - 1} \tag{14}$$

Finally, by substitution, $N_H(n)$ can be written as follows in Eq. (15):

$$N_H(n) = N_{fH}\left(1 - e^{-\frac{3n}{n_{95}}}\right) \tag{15}$$

where α is replaced by $\frac{3}{n_{95}}$ and n_{95} represents the number of rest periods needed to achieve 95% of N_{fH}.

Santagata et al. also conducted a study to explore the self-healing properties of various asphalt binders while considering the influence of steric hardening effects [26]. The authors devised a simple model to isolate the contribution of steric hardening from the overall stiffness gain observed during rest periods, thereby emphasizing the recovery in mechanical properties solely attributable to self-healing. The investigation encompassed three distinct binders, and the healing test procedure involved applying a continuous sinusoidal torque in stress-controlled mode. The first step of the tests consists of a preliminary conditioning phase in which a low oscillatory shear strain of 0.01% was applied to the specimen at a frequency of 10 Hz, and the evolution of the complex modulus over time was consequently monitored. An intermediate rest period was introduced between the first (loading) and second (reloading) oscillation phases. Using this approach, they were able to evaluate the true impact of self-healing, independent of any influence from steric hardening. In both the loading and reloading phases, all binders were subjected to a constant oscillatory shear stress amplitude of 250 kPa with a frequency of 10 Hz. Throughout each rest period, they employed a low oscillatory shear strain (0.01%) that had been applied during the conditioning phase to track the time-dependent evolution of the complex modulus. A no-damage condition was also considered in the binder testing. In this case, the materials were subjected to steric hardening tests, in which a prolonged rest period was imposed (with 0.01% shear strain at 10 Hz) following the initial thermal conditioning phase to evaluate isothermal growth of complex modulus due to molecular rearrangement. The assessment of the binders' self-healing potential is based on a comparative analysis of the time-dependent increase in the complex modulus, ΔG*(t), observed during rest periods initiated at various damage levels. The $\Delta G^*(t)$ values obtained from both healing and steric hardening tests are plotted for all the binders under consideration (Fig. 5). $\Delta G^*(t)$ represents the difference between the complex modulus measured at time t during the rest period and the complex modulus recorded at the beginning of the rest period ($G^*(t) - G^*_{iR}$). Results show that a short-term instantaneous healing effect occurs, which the authors attribute to the work of cohesion between the fracture faces of the asphalt binder. Consequently, long-term healing occurs more gradually and is governed by the molecular diffusion properties of the material.

An analytical approach was used to model experimental data, leading to the following expression (Eq. 16):

$$\Delta G^*(t) = \Delta G_\infty \left(1 - \frac{1}{e^{\alpha t}}\right)^\beta \tag{16}$$

where α and β are non-linear parameters and ΔG_∞ is the asymptotic value of $\Delta G^*(t)$ (representing the maximum gain in stiffness that can virtually be reached after an infinite rest time).

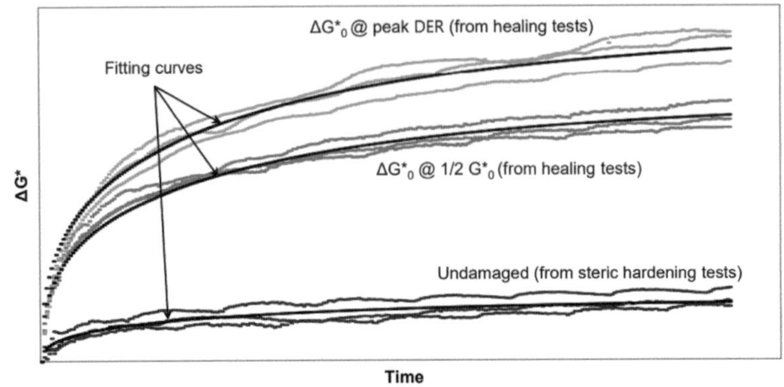

Fig. 5 Complex modulus' evolution during rest periods following various degrees of damage [26]

Researchers subtracted the asymptotic increments of the complex modulus obtained from steric hardening tests from those derived from healing tests. This subtraction yielded the differences represented as $\Delta G'_{\infty}$, which indicates the theoretical maximum stiffness gains achievable by the binders solely through self-healing contributions.

For an ideal material exhibiting a full self-healing capability, $\Delta G'_{\infty}$, and $G^*_{fC} - G^*_{iR}$ should be equal (G^*_{fC} being the value obtained after the first preliminary conditioning phase). In simpler terms, if a material is provided with a sufficiently long rest period, it would have the capability to fully recover from the damage caused before load interruption. By plotting $\Delta G'_{\infty}$, as a function of $G^*_{fC} - G^*_{iR}$ (Fig. 6), the authors demonstrate that all data points lie below the equality line, indicating that the maximum stiffness gain achievable with infinite rest time is consistently lower than the actual stiffness loss experienced by each binder. This finding suggests that the total damage experienced by the material comprises two distinct components: reversible damage, which can be progressively recovered during rest, and non-reversible damage caused by the formation of internal micro-fractures that cannot be healed.

The researchers also quantified the extent of reversible damage by introducing a relative reversible damage index (RRD) in the analysis. This index offers a quantitative measure of the true self-healing potential of the binders, expressed in relative terms (Eq. 17):

$$RDD = \frac{\Delta G'_{\infty}}{G^*_{fC} - G^*_{iR}} .100 \tag{17}$$

In a similar approach aimed at distinguishing the effects of self-healing, heating, and steric effects, Kim and Roque developed a fracture mechanics model for hot mix asphalt [4]. The model aimed to identify and quantify the healing capability of the asphalt mixture based on the resilient modulus measured during rest periods. To achieve this, they conducted indirect tensile strength tests, subjecting the mixture to 1000 cyclic loads with a haversine load duration of 0.1 s, followed by rest periods

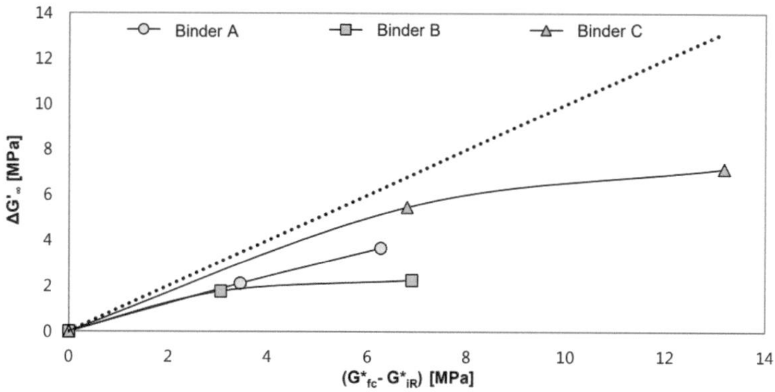

Fig. 6 Comparison between the theoretical maximum stiffness gain and the actual stiffness loss due to loading [26]

of 0.9 s. Their findings revealed a direct proportionality between total healing and damage recovery rates. Furthermore, they introduced a new method for measuring the damage recovery rate, specifically excluding the stiffness reduction caused by heating and steric softening.

To distinguish between proper self-healing and simple recovery of reversible material properties, a specifically conceived test and analysis approach was developed at ENTPE [1–3, 27] for asphalt mixtures. The tensile-compression test protocol allows isolating and quantifying biasing effects (non-linearity, heating, and thixotropy) and unrecovered variations of thermomechanical properties occurring during laboratory fatigue tests with rest periods. The testing procedure is divided into two parts:

1. The first part consists of a series of complex modulus tests performed on each tested sample in undamaged conditions at imposed temperatures and deformation levels, respectively ranging from 8 to 14 °C and from approximately 20–110 μm/m. The results obtained allow the evaluation of the variations of norm and phase angle of complex modulus with temperature and applied strain amplitude using regression equations.
2. In the second part of the test procedure, five fatigue loops of 100,000 cycles each are performed at an imposed deformation level of 100 μm/m. After each fatigue period, a rest period of 48 h is maintained. During this part of the procedure, the temperature of the thermal chamber is always kept at 10 °C. During rest periods, brief complex modulus tests are conducted at regular intervals to track the evolution of mechanical properties. Surface and internal temperatures of the sample are measured throughout the test to account for the heating phenomenon due to viscous energy dissipation.

The effects of non-linearity and self-heating on variations of complex modulus during cyclic loading and rest are estimated using the regression equations derived from the first part of the test procedure. The rest of the recovery of complex modulus during each 48-h rest period is attributed to the effect of thixotropy. For each of

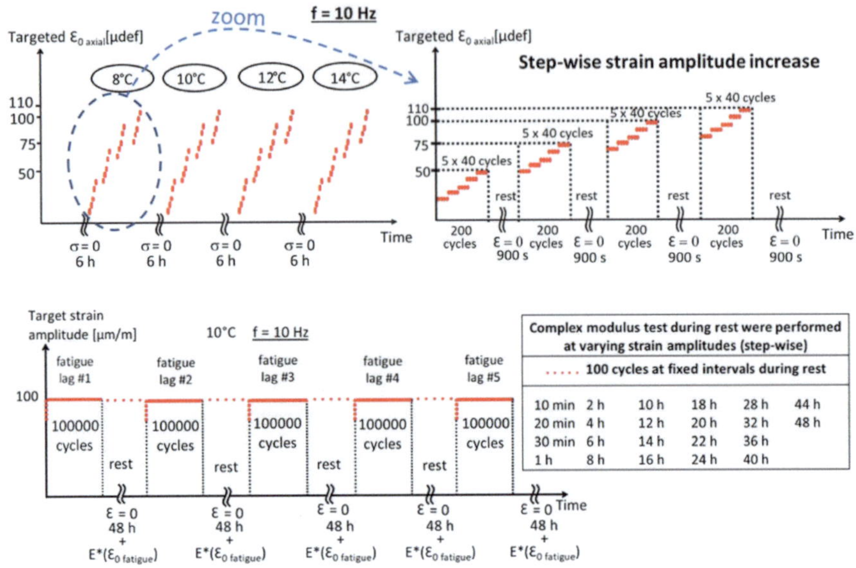

Fig. 7 Scheme of the test protocol developed at ENTPE to quantify recovery and unrecovered variations of material properties during fatigue and rest tests on asphalt mixtures: (top) complex modulus tests at different temperatures and strain amplitudes in undamaged conditions: (bottom) fatigue (100,000 cycles) and rest (48 h) intervals, with short complex modulus tests performed during rest [3]

the five fatigue and rest intervals, unrecovered variations in the complex modulus are directly observed. A scheme of the test protocol and an example of the results obtained are shown in Figs. 7 and 8, respectively.

The main conclusions obtained by the different studies carried out are as follows:

- The estimation of biasing effects shows no significant differences according to the VEL parameter used for analysis (|E*| or φ).
- At least 90% of the total variations observed during each fatigue loop (considered separately) are completely reversible. This finding reinforces the need to distinguish between the restoration of material properties after actual damage and the simple recovery of reversible variations of material properties when studying the self-healing of asphalt materials.
- Among the recovery of mechanical properties, the main contributions are due to non-linearity and thixotropy. Self-heating has a remarkably less significant effect (approximately equal to 10% of the total variations).

In a different approach, some authors have presented numerical methods for modeling the self-healing behavior of asphalt mixtures. These methods provide a powerful and versatile approach to understanding the complex mechanisms involved in the healing process of asphalt materials, allowing the simulation and analysis of the healing phenomena at different scales. Qiu et al. opted to model the cracking and

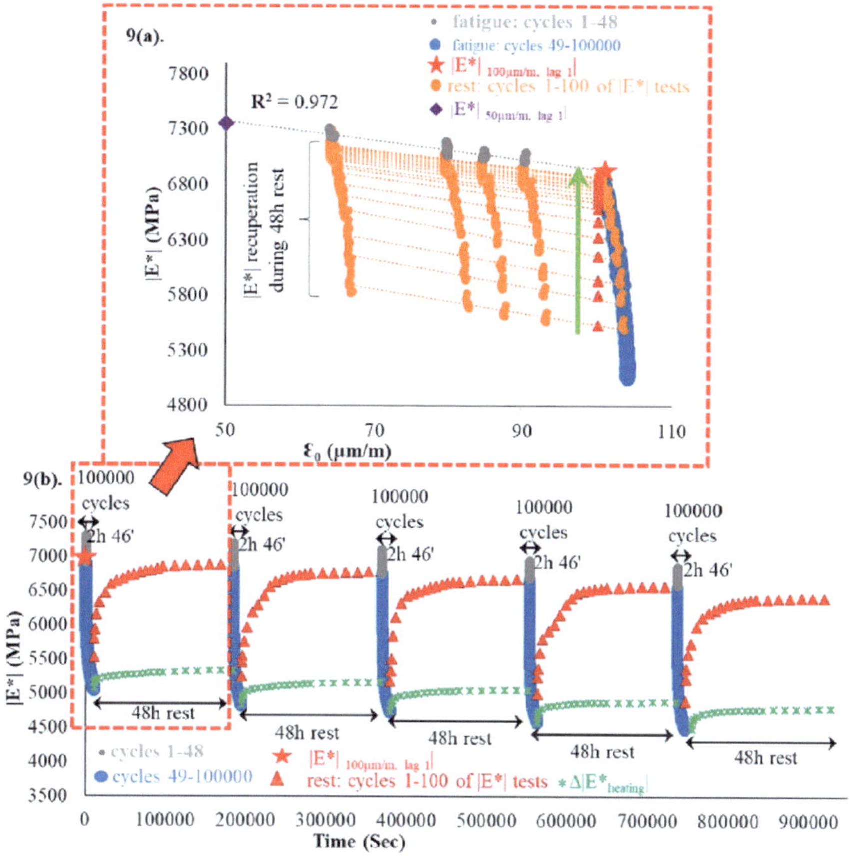

Fig. 8 Example of results obtained during fatigue and rest periods for an asphalt mixture: **a** evolution of norm of complex modulus during the first fatigue and rest (from short complex modulus tests) lag as a function of applied strain amplitude; **b** results obtained during the five fatigue and rest intervals on the same sample [3]

healing behaviors of asphalt mixtures utilizing the smeared crack Cohesive Zone Model (CZM) approach within a commercially available Finite Element Method (FEM) code called FEMMASSE [28]. The cohesive zone model is used to simulate the cracking and healing behavior in materials. The functioning of the model involves a five-step process: (i) Definition of the cohesive zone, (ii) Traction–separation behavior, (iii) Activation, (iv) Post-peak softening, and (v) Healing.

(i) Definition of the cohesive zone: In the CZM, a cohesive zone is defined ahead of the crack tip. This zone represents a region where cohesive forces between material particles are active. When the material is damaged, the crack propagates within this cohesive zone.
(ii) Traction–separation behavior: The cohesive zone is characterized by a traction–separation relationship, which describes how the traction (cohesive force)

between particles varies with the separation (crack opening) between them. This relationship can be either intrinsic, where a penalty stiffness is used for pre-peak stiffness until the maximum traction is reached, and then a descending slope of the traction–separation relationship is applied, or extrinsic, where the penalty stiffness is replaced by the stress–strain relationship for the undamaged material until the maximum traction is reached.

(iii) Activation of the model: When the applied local stress exceeds the maximum traction force of the cohesive zone, the CZM becomes active. This means that cracking and crack propagation begin to be simulated.

(iv) Post-peak softening: After the peak stress, the material undergoes post-peak softening, which means that the traction decreases as the crack propagates further. This behavior is determined by the specified traction–separation relationship in the model.

(v) Healing: The concept of healing in the CZM assumes that under certain conditions, such as resting period, temperature, compressive stress, etc., the material can heal. The healing process is considered a reverse cracking process, where the traction between particles increases, and the crack partially or completely closes.

Experimental tests were performed to assess the model using a notched beam fully glued on a low-modulus rubber foundation. A symmetric monotonic load was applied with loading unloading-healing-reloading cycles. This setup, represented in Fig. 9, allows a full closure of the crack with the help of the rubber foundation, which guarantees an autonomous healing process under different crack phases.

All components except for the asphalt mixture were assigned elastic material properties, which were modeled with a viscoelastic behavior. The viscoelastic properties of the asphalt mixtures were represented using a Generalized Maxwell Model. A finite element model was developed in FEMMASSE to simulate test results using the CZM approach (Fig. 10).

The findings demonstrate that the CZM model integrated into the FEM software successfully simulated the cracking and healing behavior. The healing process was

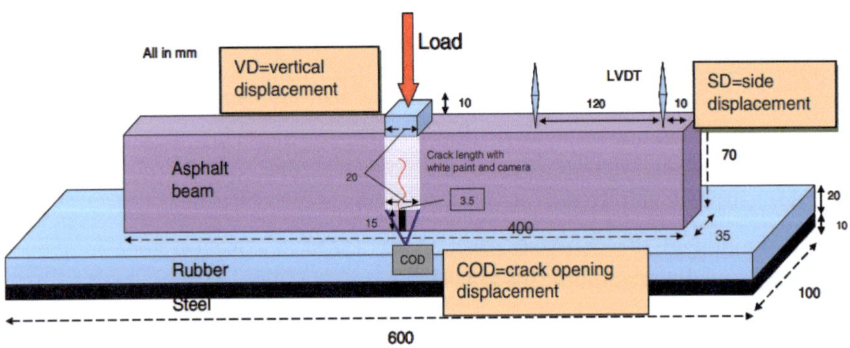

Fig. 9 Scheme of the experimental setup developed by Qiu et al. [7]

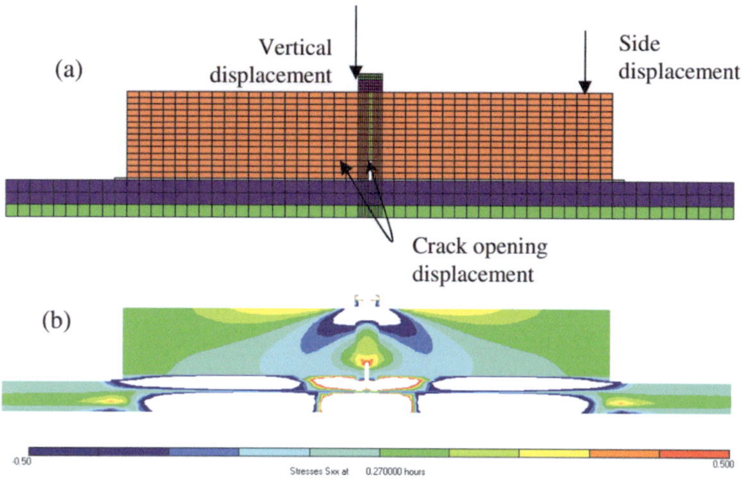

Fig. 10 Results of numerical simulations made using a finite element model developed in FEMMASSE [28]

directly associated with reduced crack length, indicating an effective crack-repairing mechanism.

In their research study, Magnanimo et al. investigated a method for modeling the cohesive, sintering, and self-healing properties of particulate materials [29]. The mechanical behavior of all particles within the material over time was simulated using a Discrete Element Method (DEM) sintering contact model. This contact model replicates the physical interactions between particles, considering dissipative, elastoplastic, adhesive, and frictional effects and pressure dependency.

The model is characterized by three "stiffness" parameters (k) that represent three significant physical effects at the contact points: (i) elasticity, (ii) plastic deformations, and (iii) contact adhesion. Additionally, the model incorporates a non-linear contact stiffness (iv). During the experimental campaign aimed at qualitatively comparing the simulations (since numerical parameters were not calibrated based on experimental results), the authors conducted uniaxial unconfined compression tests on compacted samples of dense asphalt mixtures. These tests were performed in displacement control mode with constant axial deformation rates. The mixtures were continuously graded, resembling the particle distribution in the numerical sample. The stability of such mixtures relies on the arrangement of particles and the cohesion provided by the asphalt binder. The results of the study show a method for modelling induced self-healing mechanisms in asphalt mixtures. The self-healing process occurs through an instantaneous increase in the particle contact adhesion k_t. This observed increase is the result of a resintering cycle, similar to the warming up of an asphalt mixture. This modelling approach holds promise for gaining valuable insights into self-healing behavior in asphalt materials. It could pave the way for It has the potential to lead to innovative strategies for enhancing the durability and performance of pavements.

5 Evaluation of Restoration of Material Properties Using ViscoElastic Continuum Damage Theory

5.1 *The ViscoElastic Continuum Damage Theory in Brief*

The model built on Visco-Elastic Continuum Damage (VECD) is a cracking damage characterization constitutive approach. Such a model has its roots in the work potential theory and correspondence principle, which have led to the definition of the internal state variable or damage parameter (S) and the material integrity or pseudo secant modulus (C). The relationship between these two variables is the damage characteristic curve. It represents the fundamental success of the VECD method because it is a material-dependent constitutive relationship independent of temperature and loading history. Such a curve describes the evolution of cracking damage within the material, where the reduction in stiffness caused by damage is discriminated from that related to the time-dependent viscoelastic relaxation due to the application of the elastic–viscoelastic correspondence principle [30, 31]. This principle is of primary importance because it allows the reduction of a viscoelastic problem to the elastic case by introducing pseudo-strains (ε^R), expressed in Eq. (18):

$$\varepsilon^R(t) = \frac{1}{E_R}\int_0^t E(t-\xi)\frac{d\varepsilon}{d\xi}d\xi \tag{18}$$

where ε is the true strain, $E(t)$ is the relaxation modulus, t and ξ are the time and time variables, E_R is the reference modulus (commonly assumed equal to 1). Pseudo-strains are also used for the calculation of the parameter C, determined in Eq. (19) as the ratio of the measured stress (σ) and the pseudo-strain (ε^R), in a given cycle:

$$C(S) = \frac{\sigma}{\varepsilon^R} \tag{19}$$

Based on Schapery's work potential theory [31], the microstructural changes are quantified utilizing the S variable, derived from the rate-dependent damage evolution law, defined in Eq. 20:

$$\frac{dS}{dt} = \left(-\frac{\partial W^R}{\partial S}\right)^\alpha \tag{20}$$

where α is defined as the damage evolution rate, and W^R is the pseudo-strain energy density, evaluated in Eq. (21) as a function of the applied stress and the pseudo-strain:

$$W^R = \frac{1}{2}\sigma \cdot \varepsilon^R \tag{21}$$

The combination of the previous equations has led to the final form of damage parameter, expressed in Eq. (22):

$$S_i = \sum_{i=1}^{N} \left(-\frac{\beta}{2} \cdot \left(\varepsilon^R\right)_i^2 \cdot (\Delta C_i) \right)^{\frac{\alpha}{1+\alpha}} \cdot (\Delta \xi_i)^{\frac{1}{1+\alpha}} \tag{22}$$

where ΔC_i and $\Delta \xi_i$ are the finite differences between C-values and reduced times at consecutive times, respectively, and β is the specimen-to-specimen variability coefficient.

5.2 *Evaluation of Restoration of Material Properties Using the VECD Approach*

The VECD approach has been developed to characterize fatigue damage. Since healing is defined as the partial or complete reversal of induced damage, researchers have thought the VECD should be extended to include self-healing. The simplest and most common way to express the healing potential is to consider the crucial parameters of the model, referred to as C and S. Thus, healing is typically assessed as the ratio of the pseudo stiffness or the internal state variable before and after the rest period.

Lee and Kim [32] incorporated microdamage healing within the VECD framework. Their study introduced a healing model, including the various mechanisms that allow the restoration of the material's stiffness or strength during rest periods. The model behavior is schematically displayed in Fig. 11, in terms of changes in pseudo secant modulus for a single rest and reloading period. The response related to the second loading phase is characterized by two regions: region I (AB') is the response of the healed material after the application of the rest period, whereas region II is the pseudo stiffness evolution of the previously undamaged material (B'D').

The healing model, shown in Eq. 23, includes the damage characteristic function of the virgin material $C_1(S_1)$, and introduces a healing function H:

$$\sigma = (C_1 + H)\varepsilon^R \tag{23}$$

The healing function is expressed by Eq. (24) for Region I, in which both the healed and virgin materials are damaged, and by Eq. (25) for Region II, where only the virgin material is damaged. In the above-mentioned equations, $C_2\left(S_{2,i}\right)$ is the increase in pseudo stiffness during the rest periods, $C_3\left(S_{3,i}\right)$ is the reduction in pseudo stiffness as the healed material is redamaged, $S_{B,i}^R$ and $S_{C,j}^R$ are the pseudo stiffness values corresponding to the points B' and C of the curves in Fig. 11.

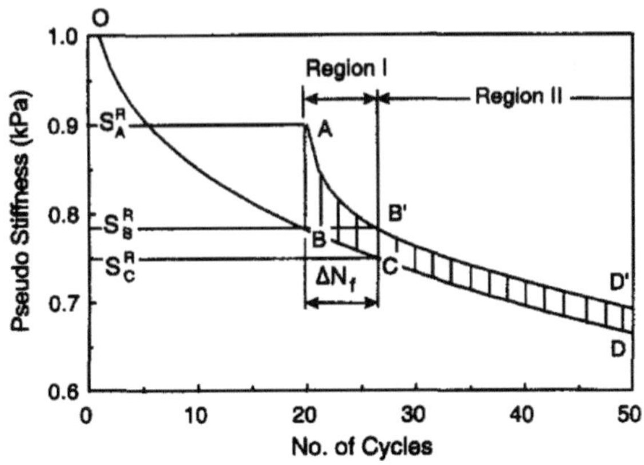

Fig. 11 Change in pseudo stiffness before and after the rest period [32]

$$H = \left[S^R_{B,i} + C_2(S_{2,i})\right]C_3(S_{3,i}) - C_1(S_1) - \sum_{j=1}^{i-1} S^R_{B,j} - S^R_{C,j} \quad (24)$$

$$H = \sum_{j=1}^{i-1} S^R_{B,j} - S^R_{C,j} \quad (25)$$

The relationships between the damage parameters, S_2 and S_3 are reported in Eqs. (26) and (27):

$$S_2 = \int_o^{t_{RP}} \left[\frac{1}{2}\left(\varepsilon^R\right)^2 \frac{\partial C_2}{\partial S_2}\right]^{\alpha_2} dt \quad (26)$$

$$S_3 = \left[-\frac{C'_A}{2}\left(\varepsilon^R\right)^2 \frac{\partial C_3}{\partial S_3}\right]^{\alpha_3} \quad (27)$$

The model, represented in Fig. 11, required a specific test method that was not clearly defined in the study.

Starting from the work of Lee and Kim, Roque and co-workers [33] proposed a simplified version of the model, including a single term for material integrity and damage in place of multiple parameters and damage functions. Moreover, such a model was thought to overcome the inaccuracies of Lee and Kim's healing model when multiple rest periods are considered. The conceptual milestone of Roque's model was the assumption that a single function exists for the damage and healing phenomena. The damaging process was thought to lead to a decrease in the material integrity along the damage curve. Instead, the healing mechanism allows the material to repair backwards along the same curve. This concept is represented in Fig. 12. The material is initially loaded from point 0 to point 1. Then, a rest period is applied, allowing healing from point 1 to point 2. Afterwards, any reloading (point 2 to point

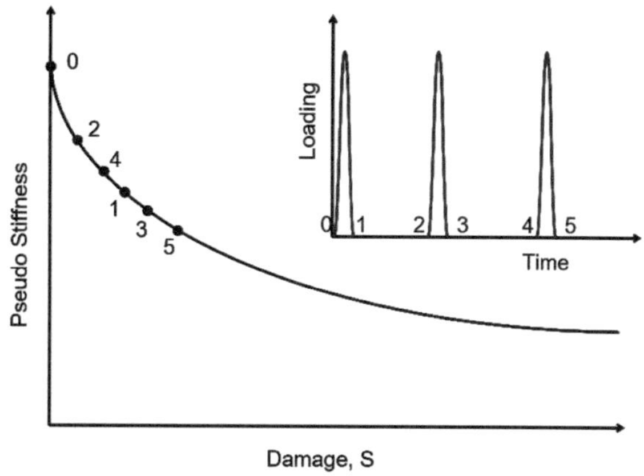

Fig. 12 Conceptual schematic diagram for the simplified model [33]

3, and point 4 to point 5) and rest period (point 3 to point 4 and so on) result in a back-and-forth movement along the same curve.

Calculation of pseudo stiffness can be conducted using Eq. (28), in which the damage reduction due to healing, S_H, requires an optimization process such that the resulting pseudo stiffness is found to be of the same value as that recorded at the end of the loading:

$$S_{i+1} = S_H + \left[-\frac{1}{2} \left(\varepsilon^R \right)^2 \frac{\partial C}{\partial S} \right]^{\alpha} \tag{28}$$

Although it presents a well-defined mathematical framework, this model lacks empirical data for validation.

Based on these pioneering works, other authors have attempted to integrate the healing effect into the VECD model using data collected from fatigue testing on asphalt binders, fine aggregate matrices, and mixtures.

Asphalt binders. The self-healing in asphalt pavement is derived from the healing potential of the asphalt binder. Therefore, the self-healing prediction of the binder phase is necessary to understand that of the entire pavement. Several attempts have been made to evaluate self-healing in binders based on the VECD [34–36].

Xie and co-workers [35] quantified the healing of asphalt binders by considering the accumulated damage growth. They successfully applied a rest-damage superposition principle to construct a healing master curve in which the rest period duration and the damage level before the rest are incorporated. Figure 13 displays the healing master curves of two different materials constructed by shifting the healing parameter evaluated in the same rest window and different damage levels.

Most recently, Miglietta and co-workers [36] developed self-healing master curves by successfully applying a rest time-rest temperature equivalency principle. Such master curves, constructed based on a non-linear viscoelastic continuum damage approach, describe the evolution of pseudo stiffness and microdamage healing in the

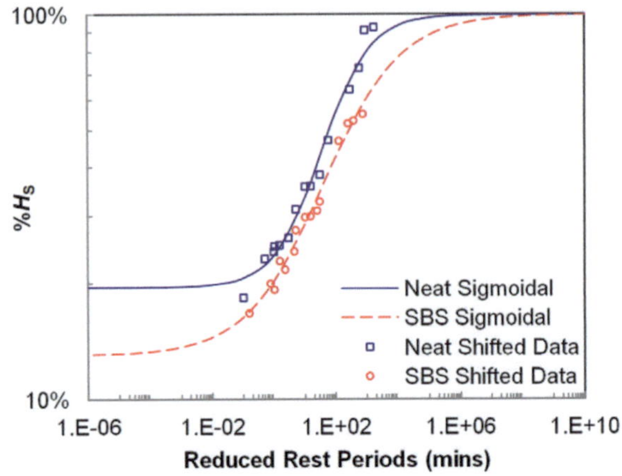

Fig. 13 Comparison of master curves for the two different asphalt binders [35]

reduced time domain. Due to the difference found in the two healing parameters, the authors speculated that the damage-based indicator was more sensitive to indicate the degree of molecular interdiffusion, representing the final steps of the healing process. Figure 14 shows an example of self-healing master curves where the curves/points corresponding to different temperatures considered for the shifting process are distinguished.

Asphalt fine aggregate matrices. When referring to the fine aggregate matrix, Palvadi and co-workers [37] quantified the material healing by developing a VECD-based model as a function of the rest period and damage level. Moving from the outcomes of Lee and Kim [32], they adopted a reduced pseudo stiffness after the rest period to evaluate the C(S) evolution of the healed material. This is schematically presented in Fig. 15, where the reduced pseudo stiffness is named C'_f. In the figure, curve C^1 describes the damage evolution in the intact matrix, whereas curve C^2 refers to partially healed material. The curve C^3 is back-predicted from the data of Region II, such that it has the same functional form as the known C^1. In conclusion, the

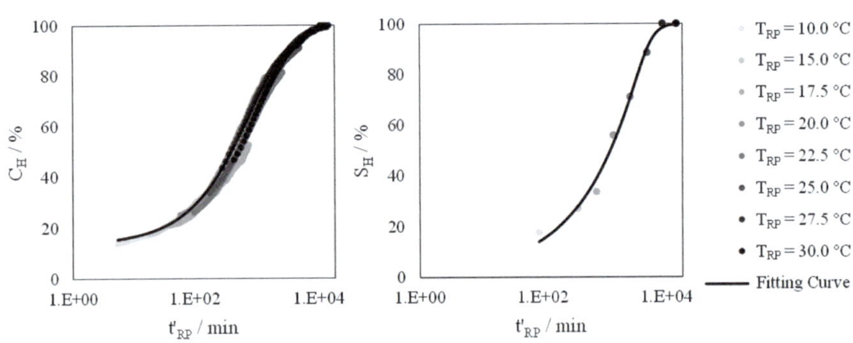

Fig. 14 Material integrity and damage healing in the reduced rest time domain [36]

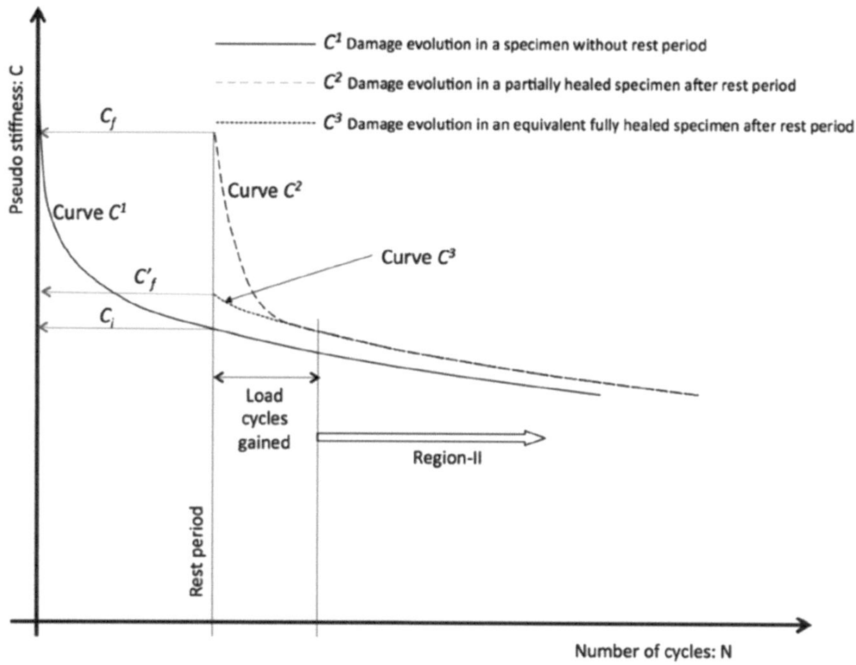

Fig. 15 Actual and modified C versus N curves are represented by Curves 2 and 3 [37]

stiffness immediately following the rest period may not be a good representation of the level of damage reversal in the matrix. Therefore, C'_f is thought to represent the equivalent effect of the partially healed and fully healed specimen.

Karki and co-workers [38] changed the abovementioned approach by proposing a method that uses the same specimen instead of several specimens to determine both damage and healing characteristics. The analytical approach adopted to define material healing is schematically illustrated in Fig. 16. The base concept of this approach is to construct the curve by shifting and combining the tail sections of time sweep curves in between the target stiffness levels.

Asphalt mixtures. In the case of asphalt mixtures, Underwood and Zeiada [39] developed a smeared-healing viscoelastic damage model to mathematically characterize mixtures with pulse-rest loading history. Such a model was compared with damage-only continuum formulations and directly used to determine the healing potential. Figure 17 shows an example of the damage characteristic curves resulting from such a smeared-healing function compared with those derived from the continuous loading case.

Later, Su and co-workers [40] adopted a smeared continuum damage-based method to capture the trend of the rest period healing effect on stiffness recovery. The final purpose was to determine the damage characteristic curve with any desired rest period duration of pulse-rest loading. For such a purpose, they used the regression

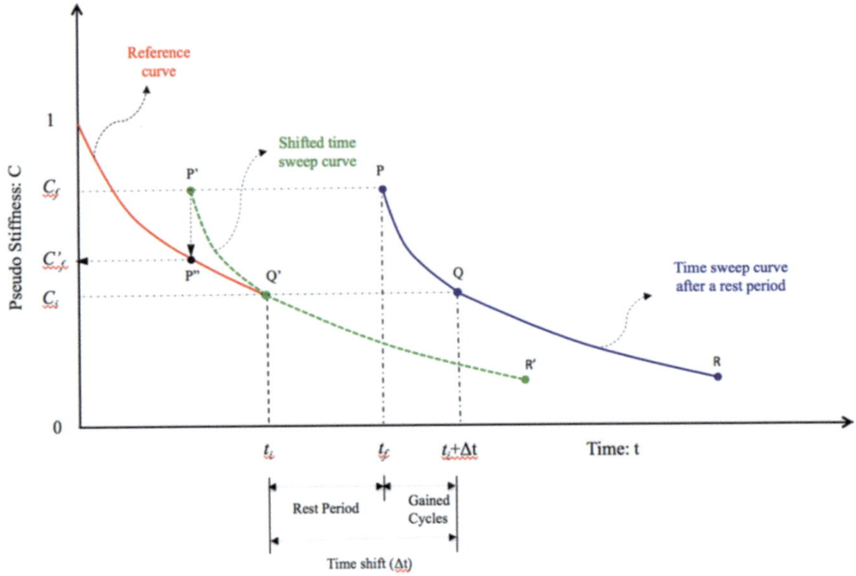

Fig. 16 Schematic illustration of shifting the time sweep curve for calculating healing potential [38]

Fig. 17 Comparison of pure damage and smeared healing damage characteristic relationships [39]

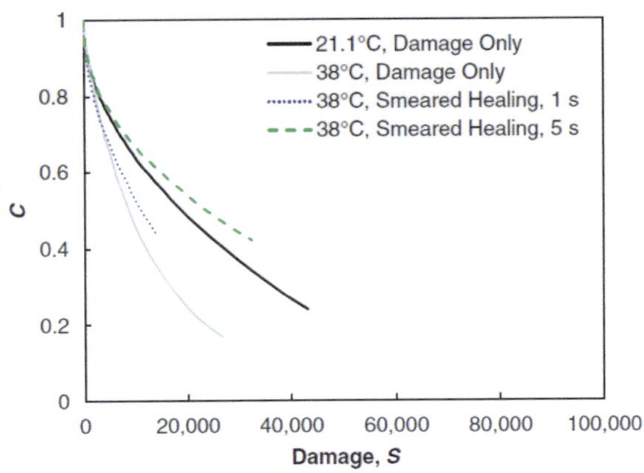

fitting technique to develop the rest period damage function. In this way, they were successful in circumventing the mechanical interpretation of the healing mechanism in the presence of numerous rest periods in pulse-rest loading cases. Figure 18 shows a set of damage characteristic curves constructed employing the rest period damage functions.

More recently, Ashouri and co-workers [41] applied the VECD theory to construct the healing master curve in the reduced rest time domain. Their proposed healing model combines the time–temperature superposition principle with a vertical shift of

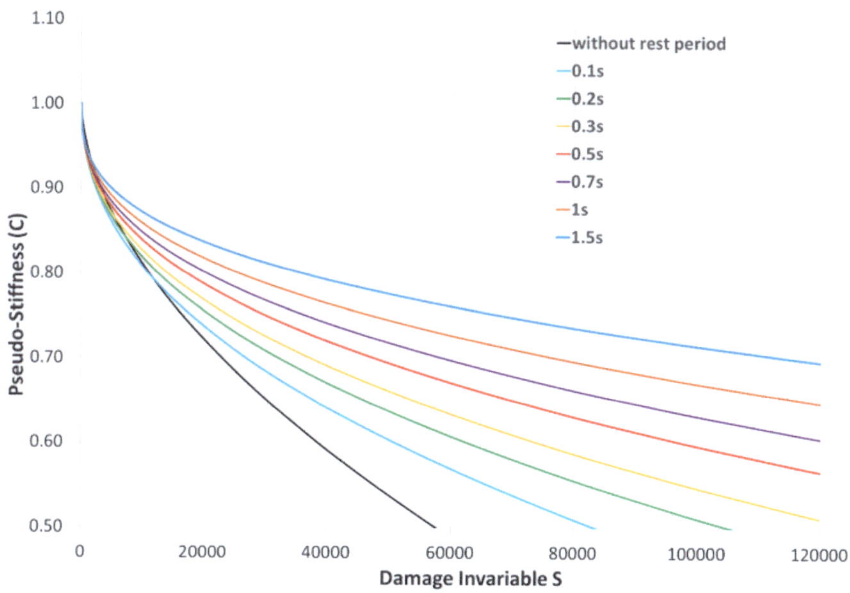

Fig. 18 Damage characteristic curves produced by rest period damage functions [40]

the master curves at different damage levels to determine a single reference healing master curve. Calibration of such a model was suggested to require three distinct samples. Results of the predicted and measured healing values calculated using such a model are provided in Fig. 19. For vertical shift, a master curve with the damage level of C = 0.7 is used as a reference healing master curve, and then the other curves are shifted toward the reference master curve.

Fig. 19 Prediction of percent healing using the healing test protocol [41]

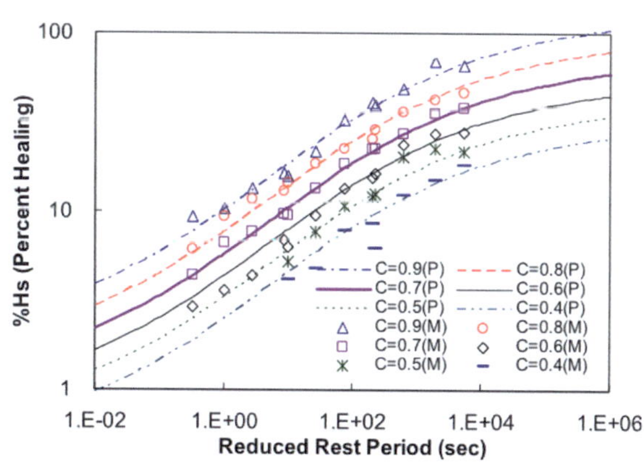

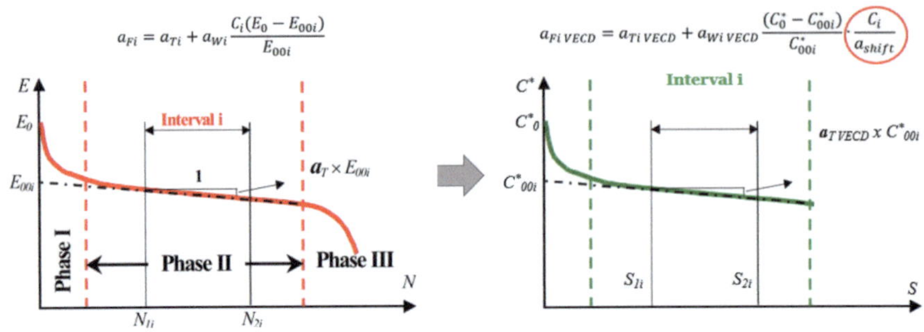

Fig. 20 Conversion from DGCB to the iVECD linearization procedure [42]

Researchers at the University of Waterloo [42] proposed an "intrinsic damage" method to evaluate the fatigue and healing behaviour of asphalt mixes. As an extension to the Simplified-VECD, the newly proposed method, called the "Intrinsic-VECD" or "iVECD" approach, corrects for the non-fatigue-related bias effects that may persist after the conversion of stresses and strains to the equivalent pseudo-variables when using the S-VECD modelling developed by Underwood and coworkers [43]. The interval-based linearization process, first developed as the DGCB method [44], was adapted to the S-VECD model to account for these bias effects as well as allowing fatigue tests to be conducted at lower strain levels compared to the recommendations in AASHTO T 342 (Standard Method of Test for Determining Dynamic Modulus of Hot Mix Asphalt (HMA)); this is especially important as bias effects have been demonstrated to have an increasing effect on the test specimen at sufficiently large strain levels [1]. Functionally, the iVECD method relies upon the same linearization method as the DGCB with a noticeable change: linearization intervals are no longer cycle-based. Instead, interval ranges are determined from the internal damage state variable (S) at the corresponding number of cycles (i.e., interval 1 ranges from S_{50k} to S_{150k}, and interval 2 ranges from S_{150k} to S_{300k}) as shown schematically in Fig. 20.

The total iVECD damage evolution rate ($a_{Ti\ VECD}$) is then determined from the linear regression of the secant pseudostiffness (C*) vs internal damage state curve. Using the slope determined in dissipated energy vs damage parameter curve ($a_{Wi\ VECD}$), the true fatigue damage rate ($a_{Fi\ VECD}$) can be determined as follows:

$$a_{FiVECD} = a_{TiVECD} + a_{WiVECD}\frac{\left(C_0^* - C_{00i}^*\right)}{C_{00i}^*}\cdot\frac{C_i}{a_{shift}} \tag{29}$$

where the time–temperature superposition (TTSP) shift factor, a_{shift}, is determined from the complex modulus master curve at the reference temperature, C_i is the non-linearity coefficient adapted from the DGCB method (i.e., $C_1 = 3/4$ from 50,000 to 150,000 cycles and for $C_2 = 2/3$ from 150,000 to 300,000 cycles), $C_0{}^*$ is the initial secant pseudostiffness, and $C_{00i}{}^*$ is the extrapolated initial secant pseudostiffness of the *ith* interval. The quotient obtained by dividing the C_i coefficient by the TTSP shift factor allows for the scaling from the number of cycles to the damage parameter, thus producing a reduced non-linearity coefficient, $C_{i,red}$. Similar to the DGCB method, correction can be applied to the secant pseudostiffness through the equation:

$$C^*_{cor} = C^*_{exp} + a_{Wi} C_{i,red} \left(C^*_0 - C^*_{00i} \right) S \tag{30}$$

where C^*_{cor} is the corrected secant pseudostiffness value, C^*_{exp} is the experimentally obtained secant pseudostiffness value at the Nth cycle. C^*_{cor} represents the initial secant pseudostiffness once the biasing effects are removed. While the iVECD approach was developed for fatigue tests, Aurilio [42] also applied the linearization method to analyse fatigue-healing tests. Figure 21 below depicts the schematic diagram of the iVECD curve for a fatigue-healing test conducted with a single intermittent rest period. The parameters of interest include:

- $C^*_{cor\_LP1}$: the C^*-intercept from the linearization of the corrected C^* values in Loading Phase 1 obtained using the equation.
- $C^*_{cor\_LP2}$: the C^*-intercept from the linearization of the corrected C^* values in Loading Phase 2 obtained using Eq. 30.
- $C^*_{fin\_LP1}$: the final secant pseudostiffness value obtained at the end of the first loading phase.
- $C^*_{ini\_LP2}$: the initial secant pseudostiffness value obtained at the beginning of the second loading phase.
- S_0: the damage parameter corresponding to the end of the first loading phase.
- ΔS: the difference obtained from the damage parameter in the second loading phase subtracted by the final secant pseudostiffness from the first loading phase, $C^*_{fin\_LP1}$.

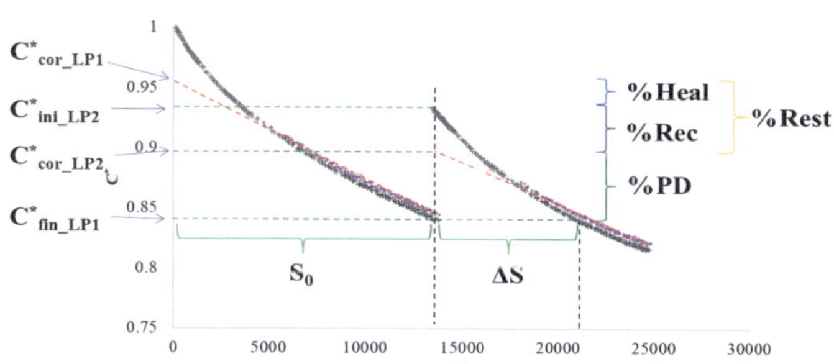

Fig. 21 Example C^*-S healing curves with several designated parameters of interest: $C^*_{cor\_LP1}$, $C^*_{cor\_LP2}$, $C^*_{ini\_LP2}$, $C^*_{fin\_LP1}$, S_0 and ΔS [42]

6 Summary

Despite significant experimental work on asphalt self-healing, comprehensive multi-scale models that describe this phenomenon remain scarce. The main difficulties lie in the lack of a precise definition of the observed phenomena behind the restoration of material properties after damage and the insufficient understanding of their fundamental underlying mechanisms. Currently, state-of-the-art research on the healing capability of asphalt materials primarily focuses on experimental mechanical characterization, with results strongly influenced by the testing techniques and conditions used.

This chapter presented an overview of the current state of research into the modeling approaches developed to simulate the self-healing capabilities of asphalt materials observed in laboratory testing. These approaches span various scales, ranging from micro-scale molecular diffusion and capillary flow models to phase-field models at intermediate scales and continuum or fracture mechanics models at the macroscopic level. Numerical simulations have also been employed to replicate results from diverse experimental tests conducted on asphalt mixtures.

Molecular diffusion and polymer-based models address microscopic processes such as drive binder flow and crack healing, highlighting how binder mobility restores material integrity. These models consider the molecular interactions that facilitate binder movement into micro-cracks, a critical factor in initiating healing. Phase-field models simulate the evolution from damaged to healed states, representing asphalt as distinct phases transitioning over time. Capillary flow models focus on the influence of binder viscosity and environmental factors, such as temperature and moisture, on binder movement and healing ability. Models based on continuum and fracture mechanics are generally used to simulate the experimental results obtained in laboratory tests on bituminous mixtures at the macroscopic level.

The review highlighted the challenges in creating a unified self-healing model for bituminous materials, including variability in materials, environmental conditions and the need for a coherent multiscale approach, based on a sound understanding of the complex phenomena occurring.

References

1. Di Benedetto, H., Nguyen, Q. T., & Sauzéat, C. (2011). Nonlinearity, heating, fatigue and thixotropy during cyclic loading of asphalt mixtures. *Road Materials and Pavement Design, 12*(1), 129–158. https://doi.org/10.1080/14680629.2011.9690356
2. Mangiafico, S., Sauzéat, C., Di Benedetto, H., Pouget, S., Olard, F., & Planque, L. (2015). Quantification of biasing effects during fatigue tests on asphalt mixes: Non-linearity, self-heating and thixotropy. *Road Materials and Pavement Design, 16*(Supl2), 73–99. https://doi.org/10.1080/14680629.2015.1077000
3. Williams, F. N., Mangiafico, S., & Sauzeat, C. (2023). Experimental evaluation of fatigue and recovery properties of a bituminous mixture during cyclic loading and rest tests. *Road Materials and Pavement Design*, 1–16. https://doi.org/10.1080/14680629.2023.2191740

4. Kim, B., & Roque, R. (2006). Evaluation of healing property of asphalt mixtures. *Transportation Research Record, 1970*(1), 84–91. https://doi.org/10.1177/0361198106197000108
5. Carpenter, S. H., & Shen, S. (2006). Dissipated energy approach to study hot-mix asphalt healing in fatigue. *Transportation Research Record, 1970*(1), 178–185.
6. Bhasin, A., Little, D. N., Bommavaram, R., & Vasconcelos, K. (2008). A framework to quantify the effect of healing in bituminous materials using material properties. *Road Materials and Pavement Design, 9*(Supl1), 219–242. https://doi.org/10.1080/14680629.2008.9690167
7. Qiu, J., Van De Ven, M., & Molenaar, A. (2013). Crack-healing investigation in bituminous materials. *Journal of Materials in Civil Engineering, 25*(7), 864–870. https://doi.org/10.1061/(ASCE)MT.1943-5533.0000744
8. Phillips, M. C. (1998). Multi-step models for fatigue and healing, and binder properties involved in healing. In *Eurobitume workshop on performance related properties for Bituminous Binders, Luxembourg*.
9. Ayar, P., Moreno-Navarro, F., & Rubio-Gámez, M. C. (2016). The healing capability of asphalt pavements: A state-of-the-art review. *Journal of Cleaner Production, 113*, 28–40. https://doi.org/10.1016/j.jclepro.2015.12.034
10. De Gennes, P. G. (1971). Reptation of a polymer chain in the presence of fixed obstacles. *The Journal of Chemical Physics, 55*(2), 572–579. https://doi.org/10.1063/1.1675789
11. Schapery, R. A. (1969). On the characterization of non-linear viscoelastic materials. *Polymer Engineering & Science, 9*(4), 295–310. https://doi.org/10.1002/pen.760090410
12. Schapery, R. A. (1989). On the mechanics of crack closing and bonding in linear viscoelastic media. *International Journal of Fracture, 39*(1–3), 163–189. https://doi.org/10.1007/BF00047448
13. Wool, R. P., & O'Connor, K. M. (1981). A theory crack healing in polymers. *Journal of Applied Physics, 52*(10), 5953–5963. https://doi.org/10.1063/1.328526
14. Bommavaram, R. R., Bhasin, A., & Little, D. N. (2009). Determining intrinsic healing properties of asphalt binders: Role of dynamic shear rheometer. *Transportation Research Record, 2126*(1), 47–54. https://doi.org/10.3141/2126-06
15. Sun, D., Lin, T., Zhu, X., Tian, Y., & Liu, F. (2016). Indices for self-healing performance assessments based on molecular dynamics simulation of asphalt binders. *Computational Materials Science, 114*, 86–93. https://doi.org/10.1016/j.commatsci.2015.12.017
16. Zhou, X., Sun, B., Wu, S., Zhang, X., Liu, Q., & Xiao, Y. (2019). Evaluation on self-healing mechanism and hydrophobic performance of asphalt modified by siloxane and polyurethane. *Journal of Wuhan University of Technology-Materials Science Edition, 34*(3), 630–637. https://doi.org/10.1007/s11595-019-2097-8
17. García, A., Norambuena-Contreras, J., Bueno, M., & Partl, M. N. (2015). Single and multiple healing of porous and dense asphalt concrete. *Journal of Intelligent Material Systems and Structures, 26*(4), 425–433. https://doi.org/10.1177/1045389X14529029
18. García, Á., Schlangen, E., Van De Ven, M., & Liu, Q. (2012). A simple model to define induction heating in asphalt mastic. *Construction and Building Materials, 31*, 38–46. https://doi.org/10.1016/j.conbuildmat.2011.12.046
19. García, A., Bueno, M., Norambuena-Contreras, J., & Partl, M. N. (2013). Induction healing of dense asphalt concrete. *Construction and Building Materials, 49*, 1–7. https://doi.org/10.1016/j.conbuildmat.2013.07.105
20. Kim, Y. R., & International Society for Asphalt Pavements (Eds.). (2014). *Asphalt pavements: Proceedings of the International Conference on Asphalt Pavements, Raleigh, North Carolina, USA, 1–5 June 2014. A Balkema books*. CRC Press.
21. Hamraoui, A., & Nylander, T. (2002). Analytical approach for the Lucas-Washburn equation. *Journal of Colloid and Interface Science, 250*(2), 415–421. https://doi.org/10.1006/jcis.2002.8288
22. Hou, Y., Wang, L., Pauli, T., & Sun, W. (2015). Investigation of the Asphalt self-healing mechanism using a phase-field model. *Journal of Materials in Civil Engineering, 27*(3), 04014118. https://doi.org/10.1061/(ASCE)MT.1943-5533.0001047

23. Kringos, N., Scarpas, A., Pauli, T., & Robertson, R. (2009). *A thermodynamic approach to healing in bitumen* (Vol. 1, pp. 123–132). Taylor and Francis Group.
24. Pauli, A. T., Robertson, R. E., Eggleston, C. M., Branthaver, J. F., & Grimes, W. (2001). Atomic force microscopy investigation of SHRP asphalts. *Preprints-American Chemical Society Division of Petroleum Chemistry, 46*(2), 104–110.
25. Canestrari, F., Virgili, A., Graziani, A., & Stimilli, A. (2015). Modeling and assessment of self-healing and thixotropy properties for modified binders. *International Journal of Fatigue, 70*, 351–360. https://doi.org/10.1016/j.ijfatigue.2014.08.004
26. Santagata, E., Baglieri, O., Tsantilis, L., & Dalmazzo, D. (2013). Evaluation of self healing properties of bituminous binders taking into account steric hardening effects. *Construction and Building Materials, 41*, 60–67. https://doi.org/10.1016/j.conbuildmat.2012.11.118
27. Mangiafico, S., Sauzeat, C., Di Benedetto, H., Pouget, S., Olard, F., & Planque, L. (2016). 'Biasing effects (non-linearity, self-heating, thixotropy) occurring during fatigue tests on bituminous mixtures', presented at the 6th Eurasphalt & Eurobitume Congress. *Prague*. https://doi.org/10.14311/EE.2016.138
28. Qiu, J., Van De Ven, M. F. C., Schlangen, E., Wu, S., & Molenaar, A. A. A. (2012). Cracking and healing modelling of asphalt mixtures. In A. Scarpas, N. Kringos, I. Al-Qadi, et al. (Eds.), *7th RILEM International Conference on Cracking in Pavements. RILEM Bookseries* (Vol. 4, pp. 1135–1144). Springer Netherlands. https://doi.org/10.1007/978-94-007-4566-7_108
29. Magnanimo, V., Ter Huerne, H. L., & Luding, S. (2012). Asphalt durability and self-healing modelling with discrete particles approach. In A. Scarpas, N. Kringos, I. Al-Qadi, et al. (Eds.), *7th RILEM International Conference on Cracking in Pavements, RILEM Bookseries* (vol. 4, pp. 1103–1114). Springer Netherlands. https://doi.org/10.1007/978-94-007-4566-7_105
30. Schapery, R. A. (1984). Correspondence principles and a generalized J integral for large deformation and fracture analysis of viscoelastic media. *International Journal of Fracture, 25*(3), 195–223. https://doi.org/10.1007/BF01140837
31. Schapery, R. A. (1990). A theory of mechanical behavior of elastic media with growing damage and other changes in structure. *Journal of the Mechanics and Physics of Solids, 38*(2), 215–253. https://doi.org/10.1016/0022-5096(90)90035-3
32. Lee, H.-J., & Kim, Y. R. (1998). Viscoelastic continuum damage model of asphalt concrete with healing. *Journal of Engineering Mechanics, 124*(11), 1224–1232. https://doi.org/10.1061/(ASCE)0733-9399(1998)124:11(1224)
33. Transportation Research Board. (2010). *Top-down cracking of hot-mix asphalt layers: Models for initiation and propagation.*
34. Baglieri, O., Tsantilis, L., & Santagata, E. (2018). Evaluation of healing potential of bituminous binders using a viscoelastic continuum damage approach. *Construction and Building Materials, 184*, 344–350. https://doi.org/10.1016/j.conbuildmat.2018.05.228
35. Xie, W., Castorena, C., Wang, C., & Richard Kim, Y. (2017). A framework to characterize the healing potential of asphalt binder using the linear amplitude sweep test. *Construction and Building Materials, 154*, 771–779. https://doi.org/10.1016/j.conbuildmat.2017.08.021
36. Fabrizio, M., Underwood, B. S., Lucia, T., Orazio, B., & Ezio, S. (2023). Self-healing master curves of bituminous binders: A non-linear viscoelastic continuum damage framework. *Road Materials and Pavement Design, 24*(sup1), 124–144. https://doi.org/10.1080/14680629.2023.2180295
37. Palvadi, S., Bhasin, A., & Little, D. N. (2012). Method to quantify healing in asphalt composites by continuum damage approach. *Transportation Research Record, 2296*(1), 86–96. https://doi.org/10.3141/2296-09
38. Karki, P., Li, R., & Bhasin, A. (2015). Quantifying overall damage and healing behaviour of asphalt materials using continuum damage approach. *International Journal of Pavement Engineering, 16*(4), 350–362. https://doi.org/10.1080/10298436.2014.942993
39. Underwood, B. S., & Zeiada, W. A. (2014). Characterization of microdamage healing in asphalt concrete with a smeared continuum damage approach. *Transportation Research Record, 2447*(1), 126–135. https://doi.org/10.3141/2447-14

40. Su, Y., Asadi, H., & Nikraz, H. (2020). VECD investigation and quantification of rest period healing within pulse-rest loading. *International Journal of Pavement Engineering, 21*(14), 1729–1745. https://doi.org/10.1080/10298436.2019.1566543
41. Ashouri, M., Wang (David), Y., Choi, Y.-T., & Kim (Richard), Y. (2021). Development of healing model and simplified characterization test procedure for asphalt concrete. *Construction and Building Materials, 271*, 121515. https://doi.org/10.1016/j.conbuildmat.2020.121515
42. Aurilio, R. M. (2023). *Investigating and developing fatigue-healing characterization of asphalt materials* [Ph.D. Thesis]. University of Waterloo.
43. Underwood, B. S., Kim, Y. R., Underwood, B. S., Baek, C., & Kim, Y. R. (2012). Simplified viscoelastic continuum damage model as platform for asphalt concrete fatigue analysis. *Transportation Research Record: Journal of the Transportation Research Board, 2296*, 36–45. https://doi.org/10.3141/2296-04
44. Baaj, H. (2002). *Comportement à la fatigue des matériaux granulaires traités aux liants hydrocarbonés* [Ph.D. Thesis]. École nationale des travaux publics de l'État.

Conclusions and Knowledge Gaps in Research in Crack-Healing of Asphalt Materials

This closing chapter synthesizes the key conclusions and findings derived from this State-of-the-Art Report. The chapter offers a critical reflection on the current state of knowledge, methodologies, and applications of self-healing technologies in asphalt pavement materials. The discussions in this chapter highlight existing research gaps, building on the interdisciplinary contributions of materials science, pavement engineering, and modelling. The chapter also defines a strategic path for future research studies to address the technical, scientific, and implementation challenges of self-healing asphalt pavements.

Self-healing in asphalt materials is a complex phenomenon involving physical, chemical, and mechanical processes. Chapter "Intrinsic Self-Healing of Asphalt Pavements" presented a detailed overview of intrinsic self-healing, focusing on the natural ability of asphalt binders to restore their mechanical integrity during rest periods. This process is primarily governed by molecular mobility within the binder phase and is influenced by temperature, time, material composition, and the scale of damage. The chapter underlined the challenges of quantifying healing effects in conventional pavement design and pointed to the limitations of current design methods that fail to fully exploit these properties. While conventional design approaches use empirical or semi-empirical factors to account for healing, a mechanistic understanding is essential to enhance prediction accuracy and material optimization.

Chapter "Extrinsic Self-Healing Technology for Asphalt Pavements" extended the discussion to extrinsic self-healing systems, including innovative technologies such as induction heating, microwave heating, encapsulated rejuvenators, and nanoparticles. These technologies are designed to actively promote healing at the earliest stage of damage, potentially extending the service life of pavements beyond conventional service life expectations. The potential environmental and economic benefits, such as reducing raw material use, lowering CO_2 emissions, and minimizing traffic disruptions, were emphasized. However, the chapter also acknowledged the technical and

© RILEM 2025

H. Baaj and O. Baglieri (eds.), *Crack-Healing of Asphalt Pavement Materials*, RILEM State-of-the-Art Reports 45, https://doi.org/10.1007/978-3-031-99443-2

logistical barriers to embedding self-healing technologies within large-scale pavement construction projects, including uniform distribution of healing agents, triggering control, and long-term durability under in-service loading and environmental conditions.

Chapter "Experimental Characterization of Self-Healing of Asphalt Materials" presented a thorough evaluation of laboratory methods for assessing healing. These include storage, intermittent, and hybrid test protocols used at multiple material scales (binder, mastic, mortar, and mixture). The chapter revealed that while several testing protocols exist, no standardized approach for evaluating healing performance exists. Additionally, the complexity of healing behaviour, particularly its dependence on test configuration, rest period duration, and induced damage mode, limits the comparability of results across different studies. A critical gap is the ability to differentiate genuine healing from confounding phenomena, such as physical relaxation or aging effects. Moreover, many test setups fail to simulate field conditions, which raises concerns regarding their applicability in performance-related specifications.

Chapter "Modeling the Crack-Healing Behavior of Asphalt Pavement Materials" focused on modelling approaches, ranging from microscale molecular diffusion and binder flow simulations to macroscale fracture and continuum mechanics. Although many models offer valuable insight into discrete aspects of healing, few can integrate the multiscale and multiphysics nature of the phenomenon. The diversity of models demonstrates a healthy academic interest, but the lack of coherence and standardization remains a fundamental issue. The absence of robust calibration datasets and model validation frameworks compounds the difficulty in transferring laboratory-based observations to predictive field-scale behaviour. Additionally, most current models do not fully account for the synergistic effects of intrinsic and extrinsic healing mechanisms or the influence of environmental cycles and loading histories typical of real-world conditions.

In light of these observations, several key research gaps emerge. First, the absence of a standardized and universally accepted protocol for testing and evaluating self-healing in asphalt binders and mixtures hampers progress in both academic and industrial settings. The variability of test outcomes due to differences in loading conditions, rest periods, and analytical approaches prevents meaningful comparisons between the different studies and undermines the reliability of the conclusions. Second, while the potential of extrinsic self-healing technologies has been demonstrated at the laboratory scale, there is insufficient evidence regarding their long-term performance under field conditions, particularly in cold climates or regions with extreme loading frequencies. Third, the complex interaction between intrinsic healing properties and externally introduced agents is poorly understood and deserves closer study, especially to understand synergistic or antagonistic effects that may arise when combining multiple healing strategies.

In terms of modelling, the main challenge lies in creating integrated, multiscale frameworks that can incorporate molecular dynamics, chemical aging, moisture diffusion, and mechanical damage evolution. This type of model would enable more realistic predictions of self-healing over the service life of the pavement. Furthermore, the incorporation of explainable AI methods and machine learning tools could support the identification of governing parameters, detect anomalies, and assist in the decision-making process. However, these approaches should not replace physics-based modelling but rather complement it by improving prediction capabilities and handling complex datasets more efficiently.

Moving forward, future research should focus on three main areas. The first is the development of rigorous, field-validated test protocols capable of isolating healing phenomena from other time- or temperature-dependent processes. The second is the design of composite self-healing systems that combine intrinsic and extrinsic approaches in a synergistic manner. This may involve multi-phase materials with tailored rheological properties or smart additives that respond to mechanical loadings. The third area involves the application of digital tools, including data analytics and numerical simulations, to develop predictive maintenance tools that can be integrated within pavement management systems, and, more generally, asset management systems.

Lastly, the real-world implementation of self-healing asphalt materials will require greater collaboration between academic researchers, industry stakeholders, and transportation agencies. Such partnerships are crucial to moving from proof-of-concept laboratory studies to more robust, cost-effective, and scalable solutions. Future efforts should also include comprehensive life-cycle assessments and cost–benefit analyses to establish the environmental and economic viability of adopting self-healing technologies at the network level. Only through a coordinated and interdisciplinary approach can the full potential of self-healing materials be realized in the context of sustainable and resilient infrastructure development.

MIX
Papier aus verantwortungsvollen Quellen
Paper from responsible sources
FSC® C105338

If you have any concerns about our products,
you can contact us on
ProductSafety@springernature.com

In case Publisher is established outside the EU,
the EU authorized representative is:
Springer Nature Customer Service Center GmbH
Europaplatz 3, 69115 Heidelberg, Germany

Printed by Libri Plureos GmbH
in Hamburg, Germany